THIRD EDITION

RESEARCH DESIGN

I dedicate this book to Karen Drumm Creswell.
She is the inspiration for my writing and my life. Because
of her, as wife, supporter, and detailed and careful editor, I am able
to work long hours and keep the home fires burning during the years that I devote
to my job and my books. Thank you from the bottom of my heart for being there for me.

THIRD EDITION

RESEARCH DESIGN

Qualitative, Quantitative, and Mixed Methods Approaches

JOHN W. CRESWELL

UNIVERSITY OF NEBRASKA–LINCOLN

Los Angeles • London • New Delhi • Singapore

For information:

SAGE Publications, Inc.
2455 Teller Road
Thousand Oaks, California 91320
E-mail: order@sagepub.com

SAGE Publications India Pvt. Ltd.
B 1/I 1 Mohan Cooperative
  Industrial Area
Mathura Road, New Delhi 110 044
India

SAGE Publications Ltd.
1 Oliver's Yard
55 City Road
London EC1Y 1SP
United Kingdom

SAGE Publications Asia-Pacific Pte. Ltd.
33 Pekin Street #02–01
Far East Square
Singapore 048763

Printed in the United States of America

Library of Congress Cataloging-in-Publication Data

Creswell, John W.
Research design: Qualitative, quantitative, and mixed methods approaches/John W. Creswell. —3rd ed.
        p. cm.
Includes bibliographical references and index.
ISBN 978-1-4129-6556-9 (cloth)
ISBN 978-1-4129-6557-6 (pbk.)

   1. Social sciences—Research—Methodology. 2. Social sciences—Statistical methods. I. Title.

H62.C6963 2009
300.72—dc22   2008006242

Printed on acid-free paper

11   12   13   14   14   13   12   11

Acquiring Editor:	Vicki Knight
Associate Editor:	Sean Connelly
Editorial Assistant:	Lauren Habib
Production Editor:	Sarah K. Quesenberry
Copy Editor:	Marilyn Power Scott
Typesetter:	C&M Digitals (P) Ltd.
Proofreader:	Marleis Roberts
Indexer:	Rick Hurd
Cover Designer:	Janet Foulger
Marketing Manager:	Stephanie Adams

Brief Contents

Detailed Contents

Analytic Contents of Research Techniques

Chapter 1. The Selection of a Research Design

- How to think about what design you should use
- Identifying a worldview with which you are most comfortable
- Defining the three types of research designs
- How to choose which one of the three designs to use

Chapter 2. Review of the Literature

- How to assess whether your topic is researchable
- The steps in conducting a literature review
- Computerized databases available for reviewing the literature
- Developing a priority for types of literature to review
- How to design a literature map
- How to write a good abstract of a research study
- Important elements of a style manual to use
- Types of terms to define
- A model for writing a literature review

Chapter 3. The Use of Theory

- The types of variables in a quantitative study
- A practical definition of a quantitative theory
- A model for writing a theoretical perspective into a quantitative study using a script
- Types of theories used in qualitative research

- Options for placing theories in a qualitative study

- How to place a theoretical lens into a mixed methods study

Chapter 4. Writing Strategies and Ethical Considerations

- Assessing how the structure of a proposal would differ depending on a qualitative, quantitative, or mixed methods design

- A writing strategy for drafting a proposal

- Developing a habit of writing

- Differences between umbrella thoughts, big thoughts, little thoughts, and attention thoughts in writing

- The hook-and-eye technique for writing consistency

- Principles of writing good prose

- Ethics issues in the research process

Chapter 5. The Introduction

- Differences among quantitative, qualitative, and mixed methods introductions

- The deficiency model for writing an introduction

- How to design a good narrative hook

- How to identify and write a research problem

- How to summarize literature about a research problem

- Distinguishing among different types of deficiencies in past literature

- Considering groups that may profit from your study

Chapter 6. The Purpose Statement

- A script for writing a qualitative purpose statement

- Considering how the script would change depending on your qualitative strategy of inquiry

- A script for writing a quantitative purpose statement

- Considering how the script would change depending on your quantitative strategy of inquiry

- A script for writing a mixed methods purpose statement
- Considering how the script would change depending on your mixed methods strategy of inquiry

Chapter 7. Research Questions and Hypotheses

- A script for writing a qualitative central question
- Considering how this script would change depending on the qualitative strategy of inquiry
- A script for writing quantitative research questions and hypotheses
- Considering how this script would change depending on the quantitative strategy of inquiry and the different types of hypotheses
- A model for presenting descriptive and inferential quantitative questions and hypotheses
- Scripts for writing different forms of research questions into a mixed methods study

Chapter 8. Quantitative Methods

- A checklist for survey research to form topic sections of a survey procedure
- Steps in analyzing data for a survey procedure
- A complete survey methods discussion
- A checklist for experimental research to form sections for an experimental procedure
- Identifying the type of experimental procedure that best fits your proposed study
- Drawing a diagram of experimental procedures
- Identifying the potential internal validity and external validity threats to your proposed study

Chapter 9. Qualitative Procedures

- A checklist for qualitative research to form topic sections of a quantitative procedure
- The basic characteristics of qualitative research

- Determining how reflexivity will be included in a proposed study
- The differences between types of data collected in qualitative research
- Distinguishing between generic forms of data analysis and analysis within strategies of inquiry
- Different levels of analysis in qualitative research
- Strategies for establishing validity for qualitative studies

Chapter 10. Mixed Methods Procedures

- Understanding a definition of mixed methods research
- How timing, weight, mixing, and theory relate to a mixed methods design
- The differences among the six models for mixed methods inquiry
- How to draw a mixed methods procedure using appropriate notation
- The different writing structures for mixed methods research

Preface

This book advances a framework, a process, and compositional approaches for designing qualitative, quantitative, and mixed methods research in the human and social sciences. Increased interest in and use of qualitative research, the emergence of mixed methods approaches, and continuing use of the traditional forms of quantitative designs have created a need for this book's unique comparison of the three approaches to inquiry. This comparison begins with preliminary consideration of philosophical assumptions for all three approaches, a review of the literature, an assessment of the use of theory in research designs, and reflections about the importance of writing and ethics in scholarly inquiry. The book then addresses the key elements of the process of research: writing an introduction, stating a purpose for the study, identifying research questions and hypotheses, and advancing methods and procedures for data collection and analysis. At each step in this process, the reader is taken through qualitative, quantitative, and mixed methods approaches.

The cover illustration depicts a mandala, a Hindu or Buddhist symbol of the universe. Creation of a mandala, much like creation of a research design, requires looking from the vantage point of a framework, an overall design, as well as focused attention on the detail—a mandala made of sand can take days to create because of the precise positioning of the pieces, which sometimes are individual grains of sand. The mandala also shows the interrelatedness of the parts of the whole, again reflecting research design, in which each element contributes to and influences the shape of a complete study.

AUDIENCE

This book is intended for graduate students and faculty who seek assistance in preparing a plan or proposal for a scholarly journal article, dissertation, or thesis. At a broader level, the book may be useful as both a reference book and a textbook for graduate courses in research methods. To best take advantage of the design features in this book, the reader needs a basic familiarity with qualitative and quantitative research; however, terms will be explained and defined and recommended strategies advanced for those needing introductory assistance in the design process. Highlighted terms in the text and a glossary of the terms at the back of the book provide a working language for understanding research. This book also is intended for a broad audience in the social and human sciences. Readers' comments since the first edition indicate that individuals using the book come from many disciplines and fields. I hope that researchers in fields such as marketing, management, criminal justice, communication studies, psychology, sociology, K–12 education, higher and postsecondary education, nursing, health sciences, urban studies, family research, and other areas will find the third edition useful.

FORMAT

In each chapter, I share examples drawn from varied disciplines. These examples are drawn from books, journal articles, dissertation proposals, and dissertations. Though my primary specialization is in education and more broadly the social sciences, the illustrations are intended to be inclusive of the social and human sciences. They reflect issues in social justice and examples of studies with marginalized individuals in our society as well as the traditional samples and populations studied by social researchers. Inclusiveness also extends to methodological pluralism in research today, and the discussion incorporates alternative philosophical ideas, diverse modes of inquiry, and numerous procedures.

This book is not a detailed method text; instead, I highlight the essential features of research design. I like to think that I have reduced research to its essential core ideas that researchers need to know to plan a thorough and thoughtful study. The coverage of research strategies of inquiry is limited to frequently used forms: surveys and experiments in quantitative research; phenomenology, ethnography, grounded theory, case studies, and narrative research in qualitative research; and concurrent, sequential, and transformative designs in mixed methods research. Although students preparing a dissertation proposal should find this book helpful, topics related to the politics of presenting and negotiating a study with graduate committees are addressed thoroughly in other texts.

Consistent with accepted conventions of scholarly writing, I have tried to eliminate any words or examples that convey a discriminatory (e.g., sexist or ethnic) orientation. Examples were selected to provide a full range of gender and cultural orientations. Favoritism also did not play into my use of qualitative and quantitative discussions: I have intentionally altered the order of qualitative and quantitative examples throughout the text. Readers should note that in the longer examples cited in this book, many references are made to other writings. Only the reference to the work I am using as an illustration will be cited, not the entire list of references embedded within any particular example. As with my earlier editions, I have maintained features to enhance the readability and understandability of the material: bullets to emphasize key points, numbered points to stress steps in a process, longer examples of complete passages with my annotations to highlight key research ideas that are being conveyed by the authors.

In this third edition of the book, new features have been added in response to developments in research and reader feedback:

● The philosophical assumptions in examining research and using theories are introduced earlier in the book as preliminary steps that researchers need to consider before they design their studies.

● The discussion about ethical issues is expanded to include more considerations related to data collection and reporting research findings.

● This edition includes, for the first time, an auxiliary CD with a complete PowerPoint slide presentation ready to use in the classroom, as well as sample activities and discussion questions.

● New Web-based technologies for literature searches are incorporated, such as Google Scholar, ProQuest, and SurveyMonkey.

● The chapter on mixed methods procedures has been extensively revised to include the latest ideas about this design. Recent articles from the Sage journal, the *Journal of Mixed Methods Research*, are included and cited.

● The second edition's chapter on definitions, limitations, and delimitations has been eliminated and the information incorporated into the chapters about reviewing the literature and the introduction to a proposal. Proposal developers today are including these ideas into other sections of a proposal.

● This third edition contains a glossary of terms that beginning and more experienced researchers can use to understand the language of research. This is especially important with the evolving language of qualitative and mixed methods research. Throughout the text, terms are carefully defined.

● I include in many chapters a delineation of research tips on different topics that have helped me advise students and faculty in research methods during the past 35 years.

● The book contains updated references throughout and attention to new editions of works.

● Features of the last edition are also maintained, such as

• The overall structure of the book with the overlays of qualitative, quantitative, and mixed methods research designs on the overall process and steps in the process of research

- The key practical strategies of understanding the philosophical assumptions of research, tips on scholarly writing, conducting a literature map of the research, scripts in writing research purpose statements and questions, and the checklists for writing detailed procedures of conducting qualitative, quantitative, and mixed methods research

- Each chapter ends with discussion questions and key references.

OUTLINE OF CHAPTERS

This book is divided into two parts. Part I consists of steps that researchers need to consider *before* they develop their proposals or plans for research. Part II discusses the various sections of a proposal.

Part I: Preliminary Considerations

This part of the book discusses preparing for the design of a scholarly study. It contains Chapters 1 through 4.

Chapter 1. The Selection of a Research Design

In this chapter, I begin by defining quantitative, qualitative, and mixed methods research and discuss them as research designs. These designs are plans for a study, and they include three important elements: philosophical assumptions, strategies of inquiry, and specific research methods. Each of these elements is discussed in detail. The choice of research design is based on considering these three elements as well as the research problem in the study, the personal experiences of the researcher, and the audiences for whom the research study will be written. This chapter should help proposal developers decide whether a qualitative, quantitative, or mixed methods design is suitable for their studies.

Chapter 2. Review of the Literature

It is important to extensively review the literature on your topic before you design your proposal. Thus you need to begin with a researchable topic and then explore the literature using the steps advanced in this chapter. This calls for setting a priority for reviewing the literature, drawing a visual map of studies that relate to your topic, writing good abstracts, employing skills learned about using style manuals, and defining key terms. This chapter should help proposal developers thoughtfully consider relevant literature on their topics and start compiling and writing literature reviews for proposals.

Chapter 3. The Use of Theory

Theories serve different purposes in the three forms of inquiry. In quantitative research, they provide a proposed explanation for the relationship among variables being tested by the investigator. In qualitative research, they may often serve as a lens for the inquiry or they may be generated during the study. In mixed methods studies, researchers employ them in many ways, including those associated with quantitative and qualitative approaches. This chapter helps proposal developers consider and plan how theory might be incorporated into their studies.

Chapter 4. Writing Strategies and Ethical Considerations

It is helpful to have an overall outline of the topics to be included in a proposal before you begin writing. Thus this chapter begins with different outlines for writing proposals; they can be used as models depending on whether your proposed study is qualitative, quantitative, or mixed methods. Then I convey several ideas about the actual writing of the proposal, such as developing a habit of writing, and grammar ideas that have been helpful to me in improving my scholarly writing. Finally, I turn to ethical issues and discuss these not as abstract ideas, but as considerations that need to be anticipated in all phases of the research process.

Part II: Designing Research

In Part II, I turn to the components of designing the research proposal. Chapters 5 through 10 address steps in this process.

Chapter 5. The Introduction

It is important to properly introduce a research study. I provide a model for writing a good scholarly introduction to your proposal. This introduction includes identifying the research problem or issue, framing this problem within the existing literature, pointing out deficiencies in the literature, and targeting the study for an audience. This chapter provides a systematic method for designing a scholarly introduction to a proposal or study.

Chapter 6. The Purpose Statement

At the beginning of research proposals, authors mention the central purpose or intent of the study. This passage is the most important statement in the entire proposal. In this chapter, you learn how to write this statement for quantitative, qualitative, and mixed methods studies, and you are provided with scripts that help you design and write these statements.

Chapter 7. Research Questions and Hypotheses

The questions and hypotheses addressed by the researcher serve to narrow and focus the purpose of the study. As another major signpost in a project, the set of research questions and hypotheses needs to be written carefully. In this chapter, the reader learns how to write both qualitative and quantitative research questions and hypotheses, as well as how to employ both forms in writing mixed methods questions and hypotheses. Numerous examples serve as scripts to illustrate these processes.

Chapter 8. Quantitative Methods

Quantitative methods involve the processes of collecting, analyzing, interpreting, and writing the results of a study. Specific methods exist in both survey and experimental research that relate to identifying a sample and population, specifying the strategy of inquiry, collecting and analyzing data, presenting the results, making an interpretation, and writing the research in a manner consistent with a survey or experimental study. In this chapter, the reader learns the specific procedures for designing survey or experimental methods that need to go into a research proposal. Checklists provided in the chapter help to ensure that all important steps are included.

Chapter 9. Qualitative Procedures

Qualitative approaches to data collection, analysis, interpretation, and report writing differ from the traditional, quantitative approaches. Purposeful sampling, collection of open-ended data, analysis of text or pictures, representation of information in figures and tables, and personal interpretation of the findings all inform qualitative procedures. This chapter advances steps in designing qualitative procedures into a research proposal, and it also includes a checklist for making sure that you cover all important procedures. Ample illustrations provide examples from phenomenology, grounded theory, ethnography, case studies, and narrative research.

Chapter 10. Mixed Methods Procedures

Mixed methods procedures employ aspects of both quantitative methods and qualitative procedures. Mixed methods research has increased in popularity in recent years, and this chapter highlights important developments in the use of this design. Six types of mixed methods designs are emphasized along with a discussion about criteria for selecting one of them based on timing, weight, mixing, and the use of theory. Figures are presented that suggest visuals that the proposal developer can design and include in a proposal. Researchers obtain an introduction to mixed methods research as practiced today and the types of designs that might be used in a research proposal.

Designing a study is a difficult and time-consuming process. This book will not necessarily make the process easier or faster, but it can provide specific skills useful in the process, knowledge about the steps involved in the process, and a practical guide to composing and writing scholarly research. Before the steps of the process unfold, I recommend that proposal developers think through their approaches to research, conduct literature reviews on their topics, develop an outline of topics to include in a proposal design, and begin anticipating potential ethical issues that may arise in the research. Part I introduces these topics.

Acknowledgments

This book could not have been written without the encouragement and ideas of the hundreds of students in the doctoral-level Proposal Development course that I have taught at the University of Nebraska-Lincoln over the years. Specific former students and editors were instrumental in its development: Dr. Sharon Hudson, Dr. Leon Cantrell, the late Nette Nelson, Dr. De Tonack, Dr. Ray Ostrander, and Diane Greenlee. Since the publication of the first edition, I have also become indebted to the students in my introductory research methods courses and to individuals who have participated in my mixed methods seminars. These courses have been my laboratories for working out ideas, incorporating new ones, and sharing my experiences as a writer and researcher. My staff in the Office of Qualitative and Mixed Methods Research at the University of Nebraska-Lincoln has also helped out extensively. I am indebted to the scholarly work of Dr. Vicki Plano Clark, Dr. Ron Shope, Dr. Kim Galt, Dr. Yun Lu, Sherry Wang, Amanda Garrett, and Alex Morales.

In addition, I am grateful for the insightful suggestions provided by the reviewers for Sage Publications. I also could not have produced this book without the support and encouragement of my friends at Sage Publications. Sage is and has been a first-rate publishing house. I especially owe much to my former editor and mentor, C. Deborah Laughton (now of Guilford Press), and to Lisa Cuevas-Shaw, Vicki Knight, and Stephanie Adams. Throughout almost 20 years of working with Sage, we have grown together to help develop research methods. Sage Publications and I gratefully acknowledge the contributions of the following reviewers:

Mahasweta M. Banerjee, University of Kansas
Miriam W. Boeri, Kennesaw State University
Sharon Anderson Dannels, The George Washington University
Sean A. Forbes, Auburn University
Alexia S. Georgakopoulos, Nova Southeastern University
Mary Enzman Hagedorn, University of Colorado at Colorado Springs
Richard D. Howard, Montana State University
Drew Ishii, Whittier College
Marilyn Lockhart, Montana State University
Carmen McCrink, Barry University

Barbara Safford, University of Northern Iowa
Stephen A. Sivo, University of Central Florida
Gayle Sulik, Vassar College
Elizabeth Thrower, University of Montevallo

About the Author

John W. Creswell is a Professor of Educational Psychology and teaches courses and writes about qualitative methodology and mixed methods research. He has been at the University of Nebraska-Lincoln for 30 years and has authored 11 books, many of which focus on research design, qualitative research, and mixed methods research and are translated into many languages and used around the world. In addition, he co-directs the Office of Qualitative and Mixed Methods Research at Nebraska that provides support for scholars incorporating qualitative and mixed methods research into projects for extramural funding. He serves as the founding coeditor for the Sage journal, *Journal of Mixed Methods Research*, and he has been an Adjunct Professor of Family Medicine at the University of Michigan and assisted investigators in the health sciences on the research methodology for their projects. He has recently been selected to be a Senior Fulbright Scholar and will be working in South Africa in October, 2008, bringing mixed methods to social scientists and to developers of documentaries about AIDS victims and families. He plays the piano, writes poetry, and actively engages in sports. Visit him at his Web site: www.johnwcreswell.com

PART I

Preliminary Considerations

This book is intended to help researchers develop a plan or proposal for a research study. Part I addresses several preliminary considerations that are necessary before designing a proposal or a plan for a study. These considerations relate to selecting an appropriate research design, reviewing the literature to position the proposed study within the existing literature, deciding on whether to use a theory in the study, and employing—at the outset—good writing and ethical practices.

CHAPTER ONE

The Selection of a Research Design

R esearch designs are plans and the procedures for research that span the decisions from broad assumptions to detailed methods of data collection and analysis. This plan involves several decisions, and they need not be taken in the order in which they make sense to me and the order of their presentation here. The overall decision involves which design should be used to study a topic. Informing this decision should be the worldview assumptions the researcher brings to the study; procedures of inquiry (called strategies); and specific methods of data collection, analysis, and interpretation. The selection of a research design is also based on the nature of the research problem or issue being addressed, the researchers' personal experiences, and the audiences for the study.

THE THREE TYPES OF DESIGNS

In this book, three types of designs are advanced: qualitative, quantitative, and mixed methods. Unquestionably, the three approaches are not as discrete as they first appear. Qualitative and quantitative approaches should not be viewed as polar opposites or dichotomies; instead, they represent different ends on a continuum (Newman & Benz, 1998). A study *tends* to be more qualitative than quantitative or vice versa. Mixed methods research resides in the middle of this continuum because it incorporates elements of both qualitative and quantitative approaches.

Often the distinction between qualitative and quantitative research is framed in terms of using words (qualitative) rather than numbers (quantitative), or using closed-ended questions (quantitative hypotheses) rather than open-ended questions (qualitative interview questions). A more complete way to view the gradations of differences between them is in the basic philosophical assumptions researchers bring to the study, the types of research strategies used overall in the research (e.g., quantitative experiments or qualitative case studies), and the specific methods employed in conducting these strategies (e.g., collecting data quantitatively on instruments versus collecting qualitative data through observing a setting). Moreover, there is a historical evolution to both approaches, with the quantitative approaches dominating the forms of research in the social sciences from the late 19th century up until the mid-20th century. During the latter half of the 20th century, interest in qualitative research increased and along with it, the development of mixed methods research (see Creswell, 2008, for more of this history). With this background, it should prove helpful to view definitions of these three key terms as used in this book:

● **Qualitative research** is a means for exploring and understanding the meaning individuals or groups ascribe to a social or human problem. The process of research involves emerging questions and procedures, data typically collected in the participant's setting, data analysis inductively building from particulars to general themes, and the researcher making interpretations of the meaning of the data. The final written report has a flexible structure. Those who engage in this form of inquiry support a way of looking at research that honors an inductive style, a focus on individual meaning, and the importance of rendering the complexity of a situation (adapted from Creswell, 2007).

● **Quantitative research** is a means for testing objective theories by examining the relationship among variables. These variables, in turn, can be measured, typically on instruments, so that numbered data can be analyzed using statistical procedures. The final written report has a set structure consisting of introduction, literature and theory, methods, results, and discussion (Creswell, 2008). Like qualitative researchers, those who engage in this form of inquiry have assumptions about testing theories deductively, building in protections against bias, controlling for alternative explanations, and being able to generalize and replicate the findings.

● **Mixed methods research** is an approach to inquiry that combines or associates both qualitative and quantitative forms. It involves philosophical assumptions, the use of qualitative and quantitative approaches, and the mixing of both approaches in a study. Thus, it is more than simply collecting and analyzing both kinds of data; it also involves the use of both approaches in tandem so that the overall strength of a study is greater than either qualitative or quantitative research (Creswell & Plano Clark, 2007).

These definitions have considerable information in each one of them. Throughout this book, I discuss the parts of the definitions so that their meanings become clear to you.

THREE COMPONENTS INVOLVED IN A DESIGN

Two important components in each definition are that the approach to research involves philosophical assumptions as well as distinct methods or procedures. **Research design**, which I refer to as the *plan or proposal to conduct research*, involves the intersection of philosophy, strategies of inquiry, and specific methods. A framework that I use to explain the inter-action of these three components is seen in Figure 1.1. To reiterate, in planning a study, researchers need to think through the philosophical worldview assumptions that they bring to the study, the strategy of inquiry that is related to this worldview, and the specific methods or procedures of research that translate the approach into practice.

Philosophical Worldviews

Although philosophical ideas remain largely hidden in research (Slife & Williams, 1995), they still influence the practice of research and need to be identified. I suggest that individuals preparing a research proposal or plan make explicit the larger philosophical ideas they espouse. This information will help explain why they chose qualitative, quantitative, or mixed methods

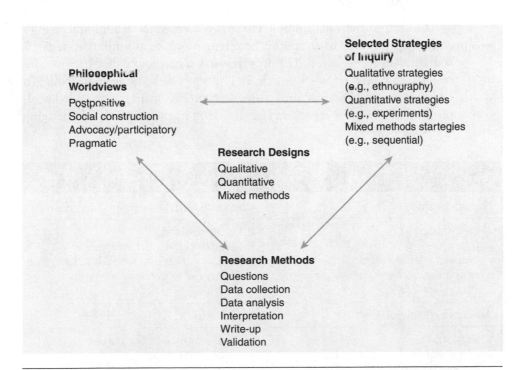

Figure 1.1   A Framework for Design—The Interconnection of Worldviews, Strategies of Inquiry, and Research Methods

approaches for their research. In writing about worldviews, a proposal might include a section that addresses the following:

● The philosophical worldview proposed in the study

● A definition of basic considerations of that worldview

● How the worldview shaped their approach to research

I have chosen to use the term **worldview** as meaning "a basic set of beliefs that guide action" (Guba, 1990, p. 17). Others have called them *paradigms* (Lincoln & Guba, 2000; Mertens, 1998); *epistemologies* and *ontologies* (Crotty, 1998), or *broadly conceived research methodologies* (Neuman, 2000). I see worldviews as a general orientation about the world and the nature of research that a researcher holds. These worldviews are shaped by the discipline area of the student, the beliefs of advisers and faculty in a student's area, and past research experiences. The types of beliefs held by individual researchers will often lead to embracing a qualitative, quantitative, or mixed methods approach in their research. Four different worldviews are discussed: postpositivism, constructivism, advocacy/participatory, and pragmatism. The major elements of each position are presented in Table 1.1.

The Postpositivist Worldview

The postpositivist assumptions have represented the traditional form of research, and these assumptions hold true more for quantitative research than qualitative research. This worldview is sometimes called the *scientific method* or doing *science research*. It is also called *positivist/postpositivist research, empirical science*, and *postpostivism*. This last term is called postpositivism because it represents the thinking after positivism, challenging

Table 1.1 Four Worldviews	
Postpositivism	**Constructivism**
• Determination • Reductionism • Empirical observation and measurement • Theory verification	• Understanding • Multiple participant meanings • Social and historical construction • Theory generation
Advocacy/Participatory	**Pragmatism**
• Political • Empowerment Issue-oriented • Collaborative • Change-oriented	• Consequences of actions • Problem-centered • Pluralistic • Real-world practice oriented

the traditional notion of the absolute truth of knowledge (Phillips & Burbules, 2000) and recognizing that we cannot be "positive" about our claims of knowledge when studying the behavior and actions of humans. The postpositivist tradition comes from 19th-century writers, such as Comte, Mill, Durkheim, Newton, and Locke (Smith, 1983), and it has been most recently articulated by writers such as Phillips and Burbules (2000).

Postpositivists hold a deterministic philosophy in which causes probably determine effects or outcomes. Thus, the problems studied by postpositivists reflect the need to identify and assess the causes that influence outcomes, such as found in experiments. It is also reductionistic in that the intent is to reduce the ideas into a small, discrete set of ideas to test, such as the variables that comprise hypotheses and research questions. The knowledge that develops through a postpositivist lens is based on careful observation and measurement of the objective reality that exists "out there" in the world. Thus, developing numeric measures of observations and studying the behavior of individuals becomes paramount for a postpositivist. Finally, there are laws or theories that govern the world, and these need to be tested or verified and refined so that we can understand the world. Thus, in the scientific method, the accepted approach to research by postpostivists, an individual begins with a theory, collects data that either supports or refutes the theory, and then makes necessary revisions before additional tests are made.

In reading Phillips and Burbules (2000), you can gain a sense of the key assumptions of this position, such as,

1. Knowledge is conjectural (and antifoundational)—absolute truth can never be found. Thus, evidence established in research is always imperfect and fallible. It is for this reason that researchers state that they do not prove a hypothesis; instead, they indicate a failure to reject the hypothesis.

2. Research is the process of making claims and then refining or abandoning some of them for other claims more strongly warranted. Most quantitative research, for example, starts with the test of a theory.

3. Data, evidence, and rational considerations shape knowledge. In practice, the researcher collects information on instruments based on measures completed by the participants or by observations recorded by the researcher.

4. Research seeks to develop relevant, true statements, ones that can serve to explain the situation of concern or that describe the causal relationships of interest. In quantitative studies, researchers advance the relationship among variables and pose this in terms of questions or hypotheses.

5. Being objective is an essential aspect of competent inquiry; researchers must examine methods and conclusions for bias. For example, standard of validity and reliability are important in quantitative research.

The Social Constructivist Worldview

Others hold a different worldview. Social constructivism (often combined with interpretivism; see Mertens, 1998) is such a perspective, and it is typically seen as an approach to qualitative research. The ideas came from Mannheim and from works such as Berger and Luekmann's (1967) *The Social Construction of Reality* and Lincoln and Guba's (1985) *Naturalistic Inquiry*. More recent writers who have summarized this position are Lincoln and Guba (2000), Schwandt (2007), Neuman (2000), and Crotty (1998), among others. **Social constructivists** hold assumptions that individuals seek understanding of the world in which they live and work. Individuals develop subjective meanings of their experiences—meanings directed toward certain objects or things. These meanings are varied and multiple, leading the researcher to look for the complexity of views rather than narrowing meanings into a few categories or ideas. The goal of the research is to rely as much as possible on the participants' views of the situation being studied. The questions become broad and general so that the participants can construct the meaning of a situation, typically forged in discussions or interactions with other persons. The more open-ended the questioning, the better, as the researcher listens carefully to what people say or do in their life settings. Often these subjective meanings are negotiated socially and historically. They are not simply imprinted on individuals but are formed through interaction with others (hence social constructivism) and through historical and cultural norms that operate in individuals' lives. Thus, constructivist researchers often address the processes of interaction among individuals. They also focus on the specific contexts in which people live and work, in order to understand the historical and cultural settings of the participants. Researchers recognize that their own backgrounds shape their interpretation, and they position themselves in the research to acknowledge how their interpretation flows from their personal, cultural, and historical experiences. The researcher's intent is to make sense of (or interpret) the meanings others have about the world. Rather than starting with a theory (as in postpostivism), inquirers generate or inductively develop a theory or pattern of meaning.

For example, in discussing constructivism, Crotty (1998) identified several assumptions:

1. Meanings are constructed by human beings as they engage with the world they are interpreting. Qualitative researchers tend to use open-ended questions so that the participants can share their views.

2. Humans engage with their world and make sense of it based on their historical and social perspectives—we are all born into a world of meaning bestowed upon us by our culture. Thus, qualitative researchers seek to understand the context or setting of the participants through visiting this context and gathering information personally. They also interpret what

they find, an interpretation shaped by the researcher's own experiences and background.

3. The basic generation of meaning is always social, arising in and out of interaction with a human community. The process of qualitative research is largely inductive, with the inquirer generating meaning from the data collected in the field.

The Advocacy and Participatory Worldview

Another group of researchers holds to the philosophical assumptions of the advocacy/participatory approach. This position arose during the 1980s and 1990s from individuals who felt that the postpositivist assumptions imposed structural laws and theories that did not fit marginalized individuals in our society or issues of social justice that needed to be addressed. This worldview is typically seen with qualitative research, but it can be a foundation for quantitative research as well. Historically, some of the advocacy/participatory (or emancipatory) writers have drawn on the works of Marx, Adorno, Marcuse, Habermas, and Freire (Neuman, 2000). Fay (1987), Heron and Reason (1997), and Kemmis and Wilkinson (1998) are more recent writers to read for this perspective. In the main, these inquirers felt that the constructivist stance did not go far enough in advocating for an action agenda to help marginalized peoples. An **advocacy/participatory worldview** holds that research inquiry needs to be intertwined with politics and a political agenda. Thus, the research contains an action agenda for reform that may change the lives of the participants, the institutions in which individuals work or live, and the researcher's life. Moreover, specific issues need to be addressed that speak to important social issues of the day, issues such as empowerment, inequality, oppression, domination, suppression, and alienation. The researcher often begins with one of these issues as the focal point of the study. This research also assumes that the inquirer will proceed collaboratively so as to not further marginalize the participants as a result of the inquiry. In this sense, the participants may help design questions, collect data, analyze information, or reap the rewards of the research. Advocacy research provides a voice for these participants, raising their consciousness or advancing an agenda for change to improve their lives. It becomes a united voice for reform and change.

This philosophical worldview focuses on the needs of groups and individuals in our society that may be marginalized or disenfranchised. Therefore, theoretical perspectives may be integrated with the philosophical assumptions that construct a picture of the issues being examined, the people to be studied, and the changes that are needed, such as feminist perspectives, racialized discourses, critical theory, queer theory, and disability theory—theoretical lens to be discussed more in Chapter 3.

Although these are diverse groups and my explanations here are generalizations, it is helpful to view the summary by Kemmis and Wilkinson (1998) of key features of the advocacy or participatory forms of inquiry:

1. Participatory action is recursive or dialectical and focused on bringing about change in practices. Thus, at the end of advocacy/participatory studies, researchers advance an action agenda for change.

2. This form of inquiry is focused on helping individuals free themselves from constraints found in the media, in language, in work procedures, and in the relationships of power in educational settings. Advocacy/participatory studies often begin with an important issue or stance about the problems in society, such as the need for empowerment.

3. It is emancipatory in that it helps unshackle people from the constraints of irrational and unjust structures that limit self-development and self-determination. The advocacy/participatory studies aim to create a political debate and discussion so that change will occur.

4. It is practical and collaborative because it is inquiry completed *with* others rather than *on* or *to* others. In this spirit, advocacy/participatory authors engage the participants as active collaborators in their inquiries.

The Pragmatic Worldview

Another position about worldviews comes from the pragmatists. Pragmatism derives from the work of Peirce, James, Mead, and Dewey (Cherryholmes, 1992). Recent writers include Rorty (1990), Murphy (1990), Patton (1990), and Cherryholmes (1992). There are many forms of this philosophy, but for many, **pragmatism** as a worldview arises out of actions, situations, and consequences rather than antecedent conditions (as in postpositivism). There is a concern with applications—what works— and solutions to problems (Patton, 1990). Instead of focusing on methods, researchers emphasize the research problem and use all approaches available to understand the problem (see Rossman & Wilson, 1985). As a philosophical underpinning for mixed methods studies, Tashakkori and Teddlie (1998), Morgan (2007), and Patton (1990) convey its importance for focusing attention on the research problem in social science research and then using pluralistic approaches to derive knowledge about the problem. Using Cherryholmes (1992), Morgan (2007), and my own views, pragmatism provides a philosophical basis for research:

● Pragmatism is not committed to any one system of philosophy and reality. This applies to mixed methods research in that inquirers draw liberally from both quantitative and qualitative assumptions when they engage in their research.

- Individual researchers have a freedom of choice. In this way, researchers are free to choose the methods, techniques, and procedures of research that best meet their needs and purposes.

- Pragmatists do not see the world as an absolute unity. In a similar way, mixed methods researchers look to many approaches for collecting and analyzing data rather than subscribing to only one way (e.g., quantitative or qualitative).

- Truth is what works at the time. It is not based in a duality between reality independent of the mind or within the mind. Thus, in mixed methods research, investigators use both quantitative and qualitative data because they work to provide the best understanding of a research problem.

- The pragmatist researchers look to the *what* and *how* to research, based on the intended consequences—where they want to go with it. Mixed methods researchers need to establish a purpose for their mixing, a rationale for the reasons why quantitative and qualitative data need to be mixed in the first place.

- Pragmatists agree that research always occurs in social, historical, political, and other contexts. In this way, mixed methods studies may include a postmodern turn, a theoretical lens that is reflective of social justice and political aims.

- Pragmatists have believed in an external world independent of the mind as well as that lodged in the mind. But they believe that we need to stop asking questions about reality and the laws of nature (Cherryholmes, 1992). "They would simply like to change the subject" (Rorty, 1983, p. xiv).

- Thus, for the mixed methods researcher, pragmatism opens the door to multiple methods, different worldviews, and different assumptions, as well as different forms of data collection and analysis.

Strategies of Inquiry

The researcher not only selects a qualitative, quantitative, or mixed methods study to conduct, the inquirer also decides on a type of study within these three choices. **Strategies of inquiry** are types of qualitative, quantitative, and mixed methods designs or models that provide specific direction for procedures in a research design. Others have called them *approaches to inquiry* (Creswell, 2007) or *research methodologies* (Mertens, 1998). The strategies available to the researcher have grown over the years as computer technology has pushed forward our data analysis and ability to analyze complex models and as individuals have articulated new procedures for conducting social science research. Select types will be emphasized in Chapters 8, 9, and 10, strategies frequently used in the social sciences. Here I introduce those that are discussed later and that are cited in examples throughout the book. An overview of these strategies is shown in Table 1.2.

Table 1.2 Alternative Strategies of Inquiry		
Quantitative	**Qualitative**	**Mixed Methods**
• Experimental designs • Non-experimental designs, such as surveys	• Narrative research • Phenomenology • Ethnographies • Grounded theory studies • Case study	• Sequential • Concurrent • Transformative

Quantitative Strategies

During the late 19th and throughout the 20th century, strategies of inquiry associated with quantitative research were those that invoked the postpositivist worldview. These include true experiments and the less rigorous experiments called *quasi-experiments* and *correlational studies* (Campbell & Stanley, 1963) and specific single-subject experiments (Cooper, Heron, & Heward, 1987; Neuman & McCormick, 1995). More recently, quantitative strategies have involved complex experiments with many variables and treatments (e.g., factorial designs and repeated measure designs). They have also included elaborate structural equation models that incorporate causal paths and the identification of the collective strength of multiple variables. In this book, I focus on two strategies of inquiry: surveys and experiments.

● **Survey research** provides a quantitative or numeric description of trends, attitudes, or opinions of a population by studying a sample of that population. It includes cross-sectional and longitudinal studies using questionnaires or structured interviews for data collection, with the intent of generalizing from a sample to a population (Babbie, 1990).

● **Experimental research** seeks to determine if a specific treatment influences an outcome. This impact is assessed by providing a specific treatment to one group and withholding it from another and then determining how both groups scored on an outcome. Experiments include true experiments, with the random assignment of subjects to treatment conditions, and quasi-experiments that use nonrandomized designs (Keppel, 1991). Included within quasi-experiments are single-subject designs.

Qualitative Strategies

In qualitative research, the numbers and types of approaches have also become more clearly visible during the 1990s and into the 21st century. Books have summarized the various types (such as the 19 strategies identified by Wolcott, 2001), and complete procedures are now available on specific qualitative inquiry approaches. For example, Clandinin and Connelly (2000) constructed a picture of what narrative researchers do.

Moustakas (1994) discussed the philosophical tenets and the procedures of the phenomenological method, and Strauss and Corbin (1990, 1998) identified the procedures of grounded theory. Wolcott (1999) summarized ethnographic procedures, and Stake (1995) suggested processes involved in case study research. In this book, illustrations are drawn from the following strategies, recognizing that approaches such as participatory action research (Kemmis & Wilkinson, 1998), discourse analysis (Cheek, 2004), and others not mentioned (see Creswell, 2007b) are also viable ways to conduct qualitative studies:

● **Ethnography** is a strategy of inquiry in which the researcher studies an intact cultural group in a natural setting over a prolonged period of time by collecting, primarily, observational and interview data (Creswell, 2007b). The research process is flexible and typically evolves contextually in response to the lived realities encountered in the field setting (LeCompte & Schensul, 1999).

● **Grounded theory** is a strategy of inquiry in which the researcher derives a general, abstract theory of a process, action, or interaction grounded in the views of participants. This process involves using multiple stages of data collection and the refinement and interrelationship of categories of information (Charmaz, 2006; Strauss and Corbin, 1990, 1998). Two primary characteristics of this design are the constant comparison of data with emerging categories and theoretical sampling of different groups to maximize the similarities and the differences of information.

● **Case studies** are a strategy of inquiry in which the researcher explores in depth a program, event, activity, process, or one or more individuals. Cases are bounded by time and activity, and researchers collect detailed information using a variety of data collection procedures over a sustained period of time (Stake, 1995).

● **Phenomenological research** is a strategy of inquiry in which the researcher identifies the essence of human experiences about a phenomenon as described by participants. Understanding the lived experiences marks phenomenology as a philosophy as well as a method, and the procedure involves studying a small number of subjects through extensive and prolonged engagement to develop patterns and relationships of meaning (Moustakas, 1994). In this process, the researcher brackets or sets aside his or her own experiences in order to understand those of the participants in the study (Nieswiadomy, 1993).

● **Narrative research** is a strategy of inquiry in which the researcher studies the lives of individuals and asks one or more individuals to provide stories about their lives. This information is then often retold or restoried by the researcher into a narrative chronology. In the end, the narrative combines views from the participant's life with those of the researcher's life in a collaborative narrative (Clandinin & Connelly, 2000).

Mixed Methods Strategies

Mixed methods strategies are less well known than either the quantitative or qualitative approaches. The concept of mixing different methods originated in 1959 when Campbell and Fisk used multimethods to study validity of psychological traits. They encouraged others to employ their multimethod matrix to examine multiple approaches to data collection. This prompted others to mix methods, and soon approaches associated with field methods, such as observations and interviews (qualitative data), were combined with traditional surveys (quantitative data; Sieber, 1973). Recognizing that all methods have limitations, researchers felt that biases inherent in any single method could neutralize or cancel the biases of other methods. Triangulating data sources—a means for seeking convergence across qualitative and quantitative methods—was born (Jick, 1979). By the early 1990s, the idea of mixing moved from seeking convergence to actually integrating or connecting the quantitative and qualitative data. For example, the results from one method can help identify participants to study or questions to ask for the other method (Tashakkori & Teddlie, 1998). Alternatively, the qualitative and quantitative data can be merged into one large database or the results used side by side to reinforce each other (e.g., qualitative quotes support statistical results; Creswell & Plano Clark, 2007). Or the methods can serve a larger, transformative purpose to advocate for marginalized groups, such as women, ethnic/racial minorities, members of gay and lesbian communities, people with disabilities, and those who are poor (Mertens, 2003).

These reasons for mixing methods have led writers from around the world to develop procedures for mixed methods strategies of inquiry, and these take the numerous terms found in the literature, such as *multimethod, convergence, integrated,* and *combined* (Creswell & Plano Clark, 2007), and shape procedures for research (Tashakkori & Teddlie, 2003).

In particular, three general strategies and several variations within them are illustrated in this book:

● **Sequential mixed methods** procedures are those in which the researcher seeks to elaborate on or expand on the findings of one method with another method. This may involve beginning with a qualitative interview for exploratory purposes and following up with a quantitative, survey method with a large sample so that the researcher can generalize results to a population. Alternatively, the study may begin with a quantitative method in which a theory or concept is tested, followed by a qualitative method involving detailed exploration with a few cases or individuals.

● **Concurrent mixed methods** procedures are those in which the researcher converges or merges quantitative and qualitative data in order to provide a comprehensive analysis of the research problem. In this design, the investigator collects both forms of data at the same time and then integrates the information in the interpretation of the overall results. Also, in this design, the researcher may embed one smaller form of data within another larger data collection in order to analyze different types of questions (the qualitative addresses the process while the quantitative, the outcomes).

● **Transformative mixed methods** procedures are those in which the researcher uses a theoretical lens (see Chapter 3) as an overarching perspective within a design that contains both quantitative and qualitative data. This lens provides a framework for topics of interest, methods for collecting data, and outcomes or changes anticipated by the study. Within this lens could be a data collection method that involves a sequential or a concurrent approach.

Research Methods

The third major element in the framework is the specific **research methods** that involve the forms of data collection, analysis, and interpretation that researchers propose for their studies. As shown in Table 1.3, it is useful to consider the full range of possibilities of data collection and to organize these methods, for example, by their degree of predetermined nature, their use of closed-ended versus open-ended questioning, and their focus on numeric versus nonnumeric data analysis. These methods will be developed further in Chapters 8 through 10.

Researchers collect data on an instrument or test (e.g., a set of questions about attitudes toward self-esteem) or gather information on a behavioral checklist (e.g., observation of a worker engaged in a complex skill). On the other end of the continuum, collecting data might involve visiting a research site and observing the behavior of individuals without predetermined questions or conducting an interview in which the individual is allowed to talk openly about a topic, largely without the use of specific

Table 1.3 Quantitative, Mixed, and Qualitative Methods		
Quantitative Methods ⟶	**Mixed Methods** ⟵	**Qualitative Methods**
• Pre-determined • Instrument based questions • Performance data, attitude data, observational data, and census data • Statistical analysis • Statistical interpretation	• Both pre-determined and emerging methods • Both open- and closed-ended questions • Multiple forms of data drawing on all possibilities • Statistical and text analysis • Across databases interpretation	• Emerging methods • Open-ended questions • Interview data, observation data, document data, and audio-visual data • Text and image analysis • Themes, patterns interpretation

questions. The choice of methods turns on whether the intent is to specify the type of information to be collected in advance of the study or allow it to emerge from participants in the project. Also, the type of data analyzed may be numeric information gathered on scales of instruments or text information recording and reporting the voice of the participants. Researchers make interpretations of the statistical results, or they interpret the themes or patterns that emerge from the data. In some forms of research, both quantitative and qualitative data are collected, analyzed, and interpreted. Instrument data may be augmented with open-ended observations, or census data may be followed by in-depth exploratory interviews. In this case of mixing methods, the researcher makes inferences across both the quantitative and qualitative databases.

RESEARCH DESIGNS AS WORLDVIEWS, STRATEGIES, AND METHODS

The worldviews, the strategies, and the methods all contribute to a research design that *tends* to be quantitative, qualitative, or mixed. Table 1.4 creates distinctions that may be useful in choosing an approach. This table also includes practices of all three approaches that are emphasized in remaining chapters of this book.

Typical scenarios of research can illustrate how these three elements combine into a research design.

- *Quantitative* approach—Postpositivist worldview, experimental strategy of inquiry, and pre- and post-test measures of attitudes

In this scenario, the researcher tests a theory by specifying narrow hypotheses and the collection of data to support or refute the hypotheses. An experimental design is used in which attitudes are assessed both before and after an experimental treatment. The data are collected on an instrument that measures attitudes, and the information is analyzed using statistical procedures and hypothesis testing.

- *Qualitative* approach—Constructivist worldview, ethnographic design, and observation of behavior

In this situation, the researcher seeks to establish the meaning of a phenomenon from the views of participants. This means identifying a culture-sharing group and studying how it develops shared patterns of behavior over time (i.e., ethnography). One of the key elements of collecting data in this way is to observe participants' behaviors by engaging in their activities.

- *Qualitative* approach—Participatory worldview, narrative design, and open-ended interviewing

For this study, the inquirer seeks to examine an issue related to oppression of individuals. To study this, stories are collected of individual oppression

Table 1.4	Qualitative, Quantitative, and Mixed Methods Approaches		
Tend to or Typically . . .	**Qualitative Approaches**	**Quantitative Approaches**	**Mixed Methods Approaches**
• Use these philosophical assumptions	• Constructivist/ advocacy/ participatory knowledge claims	• Post-positivist knowledge claims	• Pragmatic knowledge claims
• Employ these strategies of inquiry	• Phenomenology, grounded theory, ethnography, case study, and narrative	• Surveys and experiments	• Sequential, concurrent, and transformative
• Employ these methods	• Open-ended questions, emerging approaches, text or image data	• Closed-ended questions, predetermined approaches, numeric data	• Both open- and closed-ended questions, both emerging and predetermined approaches, and both quantitative and qualitative data and analysis
• Use these practices of research as the researcher	• Positions him- or herself • Collects participant meanings • Focuses on a single concept or phenomenon • Brings personal values into the study • Studies the context or setting of participants • Validates the accuracy of findings • Makes interpretations of the data • Creates an agenda for change or reform • Collaborates with the participants	• Tests or verifies theories or explanations • Identifies variables to study • Relates variables in questions or hypotheses • Uses standards of validity and reliability • Observes and measures information numerically • Uses unbiased approaches • Employs statistical procedures	• Collects both quantitative and qualitative data • Develops a rationale for mixing • Integrates the data at different stages of inquiry • Presents visual pictures of the procedures in the study • Employs the practices of both qualitative and quantitative research

using a narrative approach. Individuals are interviewed at some length to determine how they have personally experienced oppression.

● *Mixed methods* approach—Pragmatic worldview, collection of both quantitative and qualitative data sequentially

The researcher bases the inquiry on the assumption that collecting diverse types of data best provides an understanding of a research problem. The study begins with a broad survey in order to generalize results to a population and then, in a second phase, focuses on qualitative, open-ended interviews to collect detailed views from participants.

CRITERIA FOR SELECTING A RESEARCH DESIGN

Given the possibility of qualitative, quantitative, or mixed methods approaches, what factors affect a choice of one approach over another for the design of a proposal? Added to worldview, strategy, and methods would be the research problem, the personal experiences of the researcher, and the audience(s) for whom the report will be written.

The Research Problem

A research problem, more thoroughly discussed in Chapter 5, is an issue or concern that needs to be addressed (e.g., the issue of racial discrimination). Certain types of social research problems call for specific approaches. For example, if the problem calls for (a) the identification of factors that influence an outcome, (b) the utility of an intervention, or (c) understanding the best predictors of outcomes, then a quantitative approach is best. It is also the best approach to use to test a theory or explanation.

On the other hand, if a concept or phenomenon needs to be understood because little research has been done on it, then it merits a qualitative approach. Qualitative research is exploratory and is useful when the researcher does not know the important variables to examine. This type of approach may be needed because the topic is new, the topic has never been addressed with a certain sample or group of people, and existing theories do not apply with the particular sample or group under study (Morse, 1991).

A mixed methods design is useful when either the quantitative or qualitative approach by itself is inadequate to best understand a research problem or the strengths of both quantitative and qualitative research can provide the best understanding. For example, a researcher may want to both generalize the findings to a population as well as develop a detailed view of the meaning of a phenomenon or concept for individuals. In this research, the inquirer first explores generally to learn what variables to study and then studies those variables with a large sample of individuals.

Alternatively, researchers may first survey a large number of individuals and then follow up with a few participants to obtain their specific language and voices about the topic. In these situations, collecting both closed-ended quantitative data and open-ended qualitative data proves advantageous.

Personal Experiences

Researchers' own personal training and experiences also influence their choice of approach. An individual trained in technical, scientific writing, statistics, and computer statistical programs and familiar with quantitative journals in the library would most likely choose the quantitative design. On the other hand, individuals who enjoy writing in a literary way or conducting personal interviews or making up-close observations may gravitate to the qualitative approach. The mixed methods researcher is an individual familiar with both quantitative and qualitative research. This person also has the time and resources to collect both quantitative and qualitative data and has outlets for mixed methods studies, which tend to be large in scope.

Since quantitative studies are the traditional mode of research, carefully worked out procedures and rules exist for them. Researchers may be more comfortable with the highly systematic procedures of quantitative research. Also, for some individuals, it can be uncomfortable to challenge accepted approaches among some faculty by using qualitative and advocacy/participatory approaches to inquiry. On the other hand, qualitative approaches allow room to be innovative and to work more within researcher-designed frameworks. They allow more creative, literary-style writing, a form that individuals may like to use. For advocacy/participatory writers, there is undoubtedly a strong stimulus to pursue topics that are of personal interest—issues that relate to marginalized people and an interest in creating a better society for them and everyone.

For the mixed methods researcher, the project will take extra time because of the need to collect and analyze both quantitative and qualitative data. It fits a person who enjoys both the structure of quantitative research and the flexibility of qualitative inquiry.

Audience

Finally, researchers write for audiences that will accept their research. These audiences may be journal editors, journal readers, graduate committees, conference attendees, or colleagues in the field. Students should consider the approaches typically supported and used by their advisers. The experiences of these audiences with quantitative, qualitative, or mixed methods studies can shape the decision made about this choice.

SUMMARY

In planning a research project, researchers need to identify whether they will employ a qualitative, quantitative, or mixed methods design. This design is based on bringing together a worldview or assumptions about research, the specific strategies of inquiry, and research methods. Decisions about choice of a design are further influenced by the research problem or issue being studied, the personal experiences of the researcher, and the audience for whom the researcher writes.

Writing Exercises

1. Identify a research question in a journal article and discuss what design would be best to study the question and why.

2. Take a topic that you would like to study, and using the four combinations of worldviews, strategies of inquiry, and research methods in Figure 1.1, discuss a project that brings together a worldview, strategies, and methods. Identify whether this would be quantitative, qualitative, or mixed methods research.

3. What distinguishes a quantitative study from a qualitative study? Mention three characteristics.

ADDITIONAL READINGS

Cherryholmes, C. H. (1992, August-September). Notes on pragmatism and scientific realism. *Educational Researcher, 14*, 13–17.

Cleo Cherryholmes discusses pragmatism as a contrasting perspective from scientific realism. The strength of this article lies in the numerous citations of writers about pragmatism and a clarification of one version of pragmatism. Cherryholmes's version points out that pragmatism is driven by anticipated consequences, reluctance to tell a true story, and the embrace of the idea that there is an external world independent of our minds. Also included in this article are numerous references to historical and recent writers about pragmatism as a philosophical position.

Crotty, M. (1998). *The foundations of social research: Meaning and perspective in the research process.* Thousand Oaks, CA: Sage.

Michael Crotty offers a useful framework for tying together the many epistemological issues, theoretical perspectives, methodology, and methods of social research. He interrelates the four components of the research process and shows in a table a representative sampling of topics of each component. He then goes on to discuss nine different theoretical orientations in social research, such as postmodernism, feminism, critical inquiry, interpretivism, constructionism, and positivism.

Kemmis, S., & Wilkinson, M. (1998). Participatory action research and the study of practice. In B. Atweh, S. Kemmis, & P. Weeks (Eds.), *Action research in practice: Partnerships for social justice in education* (pp. 21–36). New York: Routledge.

Stephen Kemmis and Mervyn Wilkinson provide an excellent overview of participatory research. In particular, they note the six major features of this inquiry approach and then discuss how action research is practiced at the individual level, the social level, or both levels.

Guba, E. G., & Lincoln, Y. S. (2005). Paradigmatic controversies, contradictions, and emerging confluences. In N. K. Denzin & Y. S. Lincoln, *The Sage handbook of qualitative research* (3rd ed., pp. 191–215). Thousand Oaks, CA: Sage

Yvonna Lincoln and Egon Guba have provided the basic beliefs of five alternative inquiry paradigms in social science research: positivism, postpositivism, critical theory, constructivism, and participatory. These extend the earlier analysis provided in the first and second editions of the *Handbook*. Each is presented in terms of ontology (i.e., nature of reality), epistemology (i.e., how we know what we know), and methodology (i.e., the process of research). The participatory paradigm adds another alternative paradigm to those originally advanced in the first edition. After briefly presenting these five approaches, they contrast them in terms of seven issues, such as the nature of knowledge, how knowledge accumulates, and goodness or quality criteria.

Neuman, W. L. (2000). *Social research methods: Qualitative and quantitative approaches.* Boston: Allyn & Bacon.

Lawrence Neuman provides a comprehensive research method text as an introduction to social science research. Especially helpful in understanding the alternative meaning of methodology is Chapter 4, titled, "The Meanings of Methodology," in which he contrasts three methodologies—positivist social science, interpretive social science, and critical social science—in terms of eight questions (e.g., What constitutes an explanation or theory of social reality? What does good evidence or factual information look like?).

Phillips, D. C., & Burbules, N. C. (2000). *Postpositivism and educational research.* Lanham, MD: Rowman & Littlefield.

D. C. Phillips and Nicholas Burbules summarize the major ideas of postpostivist thinking. Through two chapters, "What is Postpositivism?" and "Philosophical Commitments of Postpositivist Researchers," the authors advance major ideas about postpositivism, especially those that differentiate it from positivism. These include knowing that human knowledge is conjectural rather than unchallengeable and that our warrants for knowledge can be withdrawn in light of further investigations.

CHAPTER TWO

Review of the Literature

Besides selecting a quantitative, qualitative, or mixed methods approach, the proposal designer also needs to review the literature about a topic. This literature review helps to determine whether the topic is worth studying, and it provides insight into ways in which the researcher can limit the scope to a needed area of inquiry. This chapter continues the discussion about preliminary considerations before launching into a proposal. It begins with a discussion about selecting a topic and writing this topic down so that the researcher can continually reflect on it. At this point, researchers also need to consider whether the topic *can* and *should* be researched. Then the discussion moves into the actual process of reviewing the literature, addressing the general purpose for using literature in a study and then turning to principles helpful in designing literature into qualitative, quantitative, and mixed methods studies.

THE RESEARCH TOPIC

Before considering what literature to use in a project, first identify a topic to study and reflect on whether it is practical and useful to undertake the study. The **topic** is the subject or subject matter of a proposed study, such as "faculty teaching," "organizational creativity," or "psychological stress." Describe the topic in a few words or in a short phrase. The topic becomes the central idea to learn about or to explore.

There are several ways that researchers gain some insight into their topics when they are initially planning their research (my assumption is that the topic is chosen by the researcher and not by an adviser or committee member): One way is to draft a brief title to the study. I am surprised at how often researchers fail to draft a title early in the development of their projects. In my opinion, the working or draft title becomes a major road sign in research—a tangible idea that the researcher can keep refocusing on and changing as the project goes on (see Glesne & Peshkin, 1992). I find that in my research, this topic grounds me and provides a sign of what I am studying, as well as a sign often used in conveying to others the central notion of my study. When students first provide their prospectuses of a research study to me, I ask them to supply a working title if they do not already have one on the paper.

How would this working title be written? Try completing this sentence, "My study is about. . . ." A response might be, "My study is about at-risk children in the junior high," or "My study is about helping college faculty become better researchers." At this stage in the design, frame the answer to the question so that another scholar might easily grasp the meaning of the project. A common shortcoming of beginning researchers is that they frame their study in complex and erudite language. This perspective may result from reading published articles that have undergone numerous revisions before being set in print. Good, sound research projects begin with straightforward, uncomplicated thoughts, easy to read and to understand. Think about a journal article that you have read recently. If it was easy and quick to read, it was likely written in general language that many readers could easily identify with, in a way that was straightforward and simple in overall design and conceptualization.

Wilkinson (1991) provides useful advice for creating a title: Be brief and avoid wasting words. Eliminate unnecessary words, such as "An Approach to . . . , A Study of . . . ," and so forth. Use a single title or a double title. An example of a double title would be, "An Ethnography: Understanding a Child's Perception of War." In addition to Wilkinson's thoughts, consider a title no longer than 12 words, eliminate most articles and prepositions, and make sure that it includes the focus or topic of the study.

Another strategy for topic development is to pose the topic as a brief question. What question needs to be answered in the proposed study? A researcher might ask, "What treatment is best for depression?" "What does it mean to be Arabic in U.S. society today?" "What brings people to tourist sites in the Midwest?" When drafting questions such as these, focus on the key topic in the question as the major signpost for the study. Consider how this question might be expanded later to be more descriptive of your study (see Chapters 6 and 7 on the purpose statement and research questions and hypotheses).

Actively elevating this topic to a research study calls for reflecting on whether the topic can and should be researched. A topic *can* be researched if researchers have participants willing to serve in the study. It also can be researched if investigators have resources to collect data over a sustained period of time and to analyze the information, such as available computer programs.

The question of *should* is a more complex matter. Several factors might go into this decision. Perhaps the most important are whether the topic adds to the pool of research knowledge available on the topic, replicates past studies, lifts up the voices of underrepresented groups or individuals, helps address social justice, or transforms the ideas and beliefs of the researcher.

A first step in any project is to spend considerable time in the library examining the research on a topic (strategies for effectively using the library and library resources appear later in this chapter). This point cannot be overemphasized. Beginning researchers may advance a great study that is complete in every way, such as in the clarity of research questions, the comprehensiveness of data collection, and the sophistication of statistical analysis. But the researcher may garner little support from faculty committees or conference planners because the study does not add anything new to the body of research. Ask, "How does this project contribute to the literature?" Consider how the study might address a topic that has yet to be examined, extend the discussion by incorporating new elements, or replicate (or repeat) a study in new situations or with new participants.

The issue of *should* the topic be studied also relates to whether anyone outside of the researcher's own immediate institution or area would be interested in the topic. Given a choice between a topic that might be of limited regional interest or one of national interest, I would opt for the latter because it would have wide appeal to a much broader audience. Journal editors, committee members, conference planners, and funding agencies all appreciate research that reaches a broad audience. Finally, the *should* issue also relates to the researcher's personal goals. Consider the time it takes to complete a project, revise it, and disseminate the results. All researchers should consider how the study and its heavy commitment of time will pay off in enhancing career goals, whether these goals relate to doing more research, obtaining a future position, or advancing toward a degree.

Before proceeding with a proposal or a study, one needs to weigh these factors and ask others for their reaction to a topic under consideration. Seek reactions from colleagues, noted authorities in the field, academic advisers, and faculty committee members.

THE LITERATURE REVIEW

Once the researcher identifies a topic that can and should be studied, the search can begin for related literature on the topic. The **literature review** accomplishes several purposes. It shares with the reader the results of other studies that are closely related to the one being undertaken. It relates a study to the larger, ongoing dialogue in the literature, filling in gaps and extending prior studies (Cooper, 1984; Marshall & Rossman, 2006). It provides a framework for establishing the importance of the study as well as a benchmark for comparing the results with other findings. All or some of these reasons may be the foundation for writing the scholarly literature into a study (see Miller, 1991 for a more extensive discussion of purposes for using literature in a study).

The Use of the Literature

Beyond the question of why literature is used is the additional issue of how it is used in research and proposals. It can assume various forms. My best advice is to seek the opinion of your adviser or faculty members as to how they would like to see the literature addressed. I generally recommend to my advisees that the literature review in a proposal be brief and summarize the major literature on the research problem; it does not need to be fully developed and comprehensive at this point, since faculty may ask for major changes in the study at the proposal meeting. In this model, the literature review is shorter—say 20 pages in length—and tells the reader that the student is aware of the literature on the topic and the latest writings. Another approach is to develop a detailed outline of the topics and potential references that will later be developed into an entire chapter, usually the second, titled "Literature review," which might run from 20 to 60 pages or so.

The literature review in a journal article is an abbreviated form of that found in a dissertation or master's thesis. It typically is contained in a section called "Related Literature" and follows the introduction to a study. This is the pattern for quantitative research articles in journals. For qualitative research articles, the literature review may be found in a separate section, included in the introduction, or threaded throughout the study. Regardless of the form, another consideration is how the literature might be reviewed, depending on whether a qualitative, quantitative, or mixed methods approach has been selected.

In *qualitative* research, inquirers use the literature in a manner consistent with the assumptions of learning from the participant, not prescribing the questions that need to be answered from the researcher's standpoint. One of the chief reasons for conducting a qualitative study is that the study is exploratory. This usually means that not much has been written about the topic or the population being studied, and the researcher seeks to listen to participants and build an understanding based on what is heard. However, the use of the literature in qualitative research varies considerably. In theoretically oriented studies, such as ethnographies or critical ethnographies, the literature on a cultural concept or a critical theory is introduced early in the report or proposal as an orienting framework. In grounded theory, case studies, and phenomenological studies, literature is less often used to set the stage for the study.

With an approach grounded in learning from participants and variation by type of qualitative research, there are several models for incorporating the literature review. I offer three placement locations, and it can be used in any or all of these locations. As shown in Table 2.1, the research might include the literature review in the introduction. In this placement, the literature provides a useful backdrop for the problem or issue that has led to the need for the study, such as who has been writing about it, who has studied it, and who has indicated the importance of studying the issue. This framing of the problem is, of course, contingent on available studies. One can find illustrations of this model in many qualitative studies employing different types of inquiry strategy.

Table 2.1 Using Literature in a Qualitative Study		
Use of the Literature	**Criteria**	**Examples of Suitable Strategy Types**
The literature is used to frame the problem in the introduction to the study.	There must be some literature available.	Typically, literature is used in all qualitative studies, regardless of type.
The literature is presented in a separate section as a review of the literature.	This approach is often acceptable to an audience most familiar with the traditional postpositivist approach to literature reviews.	This approach is used with those studies employing a strong theory and literature background at the beginning of a study, such as ethnographies and critical theory studies.
The literature is presented in the study at the end; it becomes a basis for comparing and contrasting findings of the qualitative study.	This approach is most suitable for the inductive process of qualitative research; the literature does not guide and direct the study but becomes an aid once patterns or categories have been identified.	This approach is used in all types of qualitative designs, but it is most popular with grounded theory, where one contrasts and compares a theory with other theories found in the literature.

A second form is to review the literature in a separate section, a model typically used in quantitative research, often found in journals with a quantitative orientation. In theory-oriented qualitative studies, such as ethnography, critical theory, or an advocacy or emancipatory aim, the inquirer might locate the theory discussion and literature in a separate section, typically toward the beginning of the write-up. Third, the researcher may incorporate the related literature in the final section, where it is used to compare and contrast with the results (or themes or categories) to emerge from the study. This model is especially popular in grounded theory studies, and I recommend it because it uses the literature inductively.

Quantitative research, on the other hand, includes a substantial amount of literature at the beginning of a study to provide direction for the research questions or hypotheses. It is also used there to introduce a problem or to describe in detail the existing literature in a section titled "Related Literature" or "Review of Literature," or some other similar phrase. Also, the literature review can introduce a theory—an explanation for expected relationships (see Chapter 3), describe the theory that will be used, and suggest why it is a useful theory to examine. At the end of a study, the literature is revisited by the researcher, and a comparison is made between the results with the existing findings in the literature. In this model, the quantitative researcher uses the literature deductively as a framework for the research questions or hypotheses.

Cooper (1984) suggests that literature reviews can be *integrative*, in which the researchers summarize broad themes in the literature. This model is popular in dissertation proposals and dissertations. A second form recommended by Cooper is a *theoretical* review in which the researcher focuses on extant theory that relates to the problem under study. This form appears in journal articles in which the author integrates the theory into the introduction. A final form suggested by Cooper is a *methodological* review, where the researcher focuses on methods and definitions. These reviews may provide both a summary of studies and a critique of the strengths and weaknesses of the methods sections. This last form is not seen frequently today in dissertations and theses.

In a *mixed methods* study, the researcher uses either a qualitative or a quantitative approach to the literature, depending on the type of strategy being used. In a sequential approach, the literature is presented in each phase in a way consistent with the method being used. For example, if the study begins with a quantitative phase, then the investigator is likely to include a substantial literature review that helps to establish a rationale for the research questions or hypotheses. If the study begins with a qualitative phase, then the literature is substantially less, and the researcher may incorporate it more into the end of the study—an inductive approach. If the researcher advances a concurrent study with an equal weight and emphasis on both qualitative and quantitative data, then the literature may take either qualitative or quantitative forms. To recap, the literature use in a mixed methods project will depend on the strategy and the relative weight given to the qualitative or quantitative research in the study.

My suggestions for using the literature in planning a qualitative, quantitative, or mixed methods study are as follows:

● In a *qualitative* study, use the literature sparingly in the beginning in order to convey an inductive design, unless the design type requires a substantial literature orientation at the outset.

● Consider the most appropriate place for the literature in a *qualitative* study, and base the decision on the audience for the project. Keep in mind the options: placing it at the beginning to frame the problem, placing it in a separate section, and using it at the end to compare and contrast with the findings.

● Use the literature in a *quantitative* study deductively, as a basis for advancing research questions or hypotheses.

● In a *quantitative* study plan, use the literature to introduce the study, describe related literature in a separate section, and to compare findings.

● If a separate review is used, consider whether the literature will be integrative summaries, theoretical reviews, or methodological reviews. A typical practice in dissertation writing is to advance an integrative review.

● In a *mixed methods* study, use the literature in a way that is consistent with the major type of strategy and the qualitative or quantitative approach most prevalent in the design.

Design Techniques

Regardless of the type of study, several steps are useful in conducting a literature review.

Steps in Conducting a Literature Review

A literature review means locating and summarizing the studies about a topic. Often these are research studies (since you are conducting a research study), but they may also include conceptual articles or thought pieces that provide frameworks for thinking about topics. There is no single way to conduct a literature review, but many scholars proceed in a systematic fashion to capture, evaluate, and summarize the literature. Here is the way I recommend:

1. Begin by identifying key words, useful in locating materials in an academic library at a college or university. These key words may emerge in identifying a topic or may result from preliminary readings.

2. With these key words in mind, next go to the library and begin searching the catalog for holdings (i.e., journals and books). Most major libraries have computerized databases, and I suggest you focus initially on journals and books related to the topic. Also, begin to search the computerized data bases that are typically reviewed by social science researchers, such as ERIC, PsycINFO, Sociofile, the Social Science Citation Index, Google Scholar, ProQuest, and others (these are reviewed later in some detail). These databases are available online using the library's Web site or they may be available on CD-ROM.

3. Initially, try to locate about 50 reports of research in articles or books related to research on your topic. Set a priority on the search for journal articles and books because they are easy to locate and obtain. Determine whether these articles and books exist in your academic library or whether you need to send for them by interlibrary loan or purchase them through a bookstore.

4. Skim this initial group of articles or chapters, and duplicate those that are central to your topic. Throughout this process, simply try to obtain a sense as to whether the article or chapter will make a useful contribution to your understanding of the literature.

5. As you identify useful literature, begin designing a **literature map** (to be discussed more fully later). This is a visual picture (or figure) of groupings of the literature on the topic, that illustrates how your particular study will contribute to the literature, positioning your own study within the larger body of research.

6. As you put together the literature map, also begin to draft summaries of the most relevant articles. These summaries are combined into the final literature review that you write for your proposal or research study. Include precise references to the literature using an appropriate style guide, such as the American Psychological Association (APA) style manual (APA, 2001) so that you have a complete reference to use at the end of the proposal or study.

7. After summarizing the literature, assemble the literature review, structuring it thematically or organizing it by important concepts. End the literature review with a summary of the major themes and suggest how your particular study further adds to the literature.

Searching Computerized Databases

To ease the process of collecting relevant material, there are some techniques useful in accessing the literature quickly through databases. **Computer databases of the literature** are now available in libraries, and they quickly provide access to thousands of journals, conference papers, and materials on many different topics. Academic libraries at major universities have purchased commercial databases as well as obtained databases in the public domain. Only a few of the major databases available will be reviewed here, but they are the major sources to journal articles and documents that you should consult to determine what literature is available on your topic.

ERIC (Educational Resources Information Center) is a free, online digital library of education research and information sponsored by the Institute of Education Sciences (IES) of the U.S. Department of Education. This database can be found at http://www.eric.ed.gov, and ERIC provides a search of 1.2 million items indexed since 1966. The collection includes journal articles, books, research syntheses, conference papers, technical reports, policy papers, and other education-related materials. ERIC indexes more than 600 journals, and links are available to full-text copies of many of the materials. To best utilize ERIC, it is important to identify appropriate descriptors for your topic, the terms used by indexers to categorize article or documents. Researchers can search through the *Thesaurus of ERIC Descriptors* (Educational Resources Information Center, 1975) or browse the online thesaurus. A **research tip** in conducting an ERIC search is to locate recent journal articles and documents on your topic. This process can be enhanced by conducting a preliminary search using descriptors from the online thesaurus and locating a journal article or document which is on your topic.

Then look closely at the descriptors used in this article and document and run another search using these terms. This procedure will maximize the possibility of obtaining a good list of articles for your literature review.

Another free database to search is Google Scholar. It provides a way to broadly search for literature across many disciplines and sources, such as peer-reviewed papers, theses, books, abstracts, and articles from academic publishers, professional societies, universities, and other scholarly organizations. The articles identified in a Google Scholar search provide links to abstracts, related articles, electronic versions of articles affiliated with a library you specify, Web searches for information about this work, and opportunities to purchase the full text of the article.

Researchers can obtain abstracts to publications in the health sciences through the free-access PubMed. This database is a service of the U.S. National Library of Medicine, and it includes over 17 million citations from MEDLINE and other life science journals for biomedical articles going back to the 1950s (www.ncbi.nlm.nih.gov). PubMed includes links to full-text articles (located in academic libraries) and other related resources. To search PubMed, the researcher uses MeSH (Medical Subject Headings) terms, the U.S. National Library of Medicine's controlled vocabulary thesaurus used for indexing articles for MEDLINE/PubMed. This MeSH terminology provides a consistent way to retrieve information about topics that may be described using different terms.

Academic libraries also have site licenses to important commercial databases. One typically available is ProQuest (http://proquest.com), which enables a researcher to search many different databases, and it is one of the largest online content repositories in the world. For example, you can search ERIC, PsycINFO, Dissertation Abstracts, Periodicals Index, Health and Medical Complete, and many more specialized databases (e.g., International Index to Black Periodicals). Because it taps into many different databases, it can be one search tool to use before using more specialized databases.

Another commercially licensed database found in many academic libraries is *Sociological Abstracts* (Cambridge Scientific Abstracts, http://www.csa.com). This database indexes over 2,000 journals, conference papers, relevant dissertation listings, book reviews, and selected books in sociology, social work, and related disciplines. For literature in the field of psychology and related areas, consult another commercial database, PsycINFO (http://www.apa.org). This database indexes 2,150 journal titles, books, and dissertations from many countries. It covers the field of psychology as well as psychological aspects of related disciplines, including medicine, psychiatry, nursing, sociology, education, pharmacology, physiology, linguistics, anthropology, business, and law. It has a Thesaurus of Psychological Index Terms to locate useful terms in a literature search.

A final commercial database available in libraries is The *Social Sciences Citation Index* (SSCI, Web of Knowledge, Thomson Scientific [http://isiweb ofknowledge.com]). It indexes 1,700 journals spanning 50 disciplines and selectively indexes relevant items from over 3,300 scientific and technical journals. It can be used to locate articles and authors who have conducted research on a topic. It is especially useful in locating studies that have referenced an important study. The SSCI enables you to trace all studies since the publication of the key study that have cited the work. Using this system, you can develop a chronological list of references that document the historical evolution of an idea or study. This chronological list can be most helpful in tracking the developing of ideas about your literature review topic.

In summary, my **research tips** for searching computer databases are to

- Use both the free, online literature databases as well as those available through your academic library.

- Search several databases, even if you feel that your topic is not strictly education, as found in ERIC, or psychology, as found in PsycINFO. Both ERIC and PsycINFO view education and psychology as broad terms for many topics.

- Use guides to terms to locate your articles, such as a thesaurus, when available.

- Locate an article that is close to your topic, then look at the terms used to describe it, and use these terms in your search.

- Use databases that provide access to full-text copies of your articles (through academic libraries or for a fee) as much as possible so that you can reduce the amount of time searching for copies of your articles.

A Priority for Selecting Literature Material

I recommend that you establish a priority in a search of the literature. What types of literature might be reviewed and in what priority? Consider the following:

1. Especially if you are examining a topic for the first time and unaware of the research on it, start with broad syntheses of the literature, such as overviews found in encyclopedias (e.g., Aikin, 1992; Keeves, 1988). You might also look for summaries of the literature on your topic presented in journal articles or abstract series (e.g., *Annual Review of Psychology, 1950–*).

2. Next, turn to journal articles in respected, national journals, especially those that report research studies. By *research*, I mean that the author or authors pose a question or hypothesis, collect data, and try to answer the question or hypothesis. There are journals widely read in your field, and typically they are publications with a high-quality editorial board consisting of individuals from around the United States or abroad. By turning to the first few pages, you can determine if an editorial board is

listed and whether it is made up of individuals from around the country or world. Start with the most recent issues of the journals and look for studies about your topic and then work backward in time. Follow up on references at the end of the articles for more sources to examine.

3. Turn to books related to the topic. Begin with research monographs that summarize the scholarly literature. Then consider entire books on a single topic by a single author or group of authors or books that contain chapters written by different authors.

4. Follow this search by recent conference papers. Look for major national conferences and the papers delivered at them. Often, conference papers report the latest research developments. Most major conferences either require or request that authors submit their papers for inclusion in computerized indices. Make contact with authors of pertinent studies. Seek them out at conferences. Write or phone them, asking if they know of studies related to your area of interest and inquire also if they have an instrument that might be used or modified for use in your study.

5. If time permits, scan the entries in *Dissertation Abstracts* (University Microfilms, 1938). Dissertations vary immensely in quality, and one needs to be selective in choosing those to review. A search of the *Abstracts* might result in one or two relevant dissertations, and you can request copies of them through interlibrary loans or through the University of Michigan Microfilm Library.

6. The Web also provides helpful materials for a literature review. The easy access and ability to capture entire articles makes this source of material attractive. However, screen these articles carefully for quality and be cautious about whether they represent rigorous, thoughtful, and systematic research suitable for use in a literature review. Online journals, on the other hand, often include articles that have undergone rigorous reviews by editorial boards. You might check to see if the journal has a refereed editorial board that reviews manuscripts and has published standards for accepting manuscripts in an editorial statement.

In summary, I place refereed journal articles high on the list because they are the easiest to locate and duplicate. They also report research about a topic. Dissertations are listed lower in priority because they vary considerably in quality and are the most difficult reading material to locate and reproduce. Caution should be used in choosing journal articles on the Web unless they are part of refereed online journals.

A Literature Map of the Research

One of the first tasks for a researcher working with a new topic is to organize the literature. As mentioned earlier, this organization enables a person to understand how the proposed study adds to, extends, or replicates research already completed.

A useful approach for this step is to design a literature map. This is an idea that I came up with several years ago, and it has been a useful tool for students to use when organizing their review of the literature for making presentations to graduate committees or summarizing the literature for a scholarly presentation or a journal article publication.

This map is a visual summary of the research that has been conducted by others, and it is typically represented in a figure. Maps are organized in different ways. One could be a hierarchical structure, with a top-down presentation of the literature, ending at the bottom with the proposed study. Another might be similar to a flowchart in which the reader understands the literature as unfolding from left to right with the farthest right-hand section advancing a proposed study. A third model might be a series of circles, with each circle representing a body of literature and the intersection of the circles the place in which the future research is indicated. I have seen examples of all of these possibilities and found them all effective.

The central idea is that the researcher begins to build a visual picture of existing research about a topic. This literature map presents an overview of existing literature. Figure 2.1 is an illustration of a map that shows the literature found on procedural justice in organizational studies (Janovec, 2001). Janovec's map illustrates a hierarchical design, and she used several principles of good map design.

- She placed her topic in the box at the top of the hierarchy.

- Next, she took the studies that she found in computer searches, located copies of these studies, and organized them into three broad subtopics (i.e., Justice Perceptions Formation, Justice Effects, and Justice in Organizational Change). For another map, the researcher may have more or fewer than four major categories, depending on the extent and publications on the topic.

- Within each box are labels that describe the nature of the studies in the box (i.e., outcomes).

- Also within each box are references to major citations illustrating its content. It is useful to use references that are current, illustrative of the topic of the box and to briefly state the references in an appropriate style, such as APA.

- Consider several levels for the literature map. In other words, major topics lead to subtopics and then to sub-subtopics.

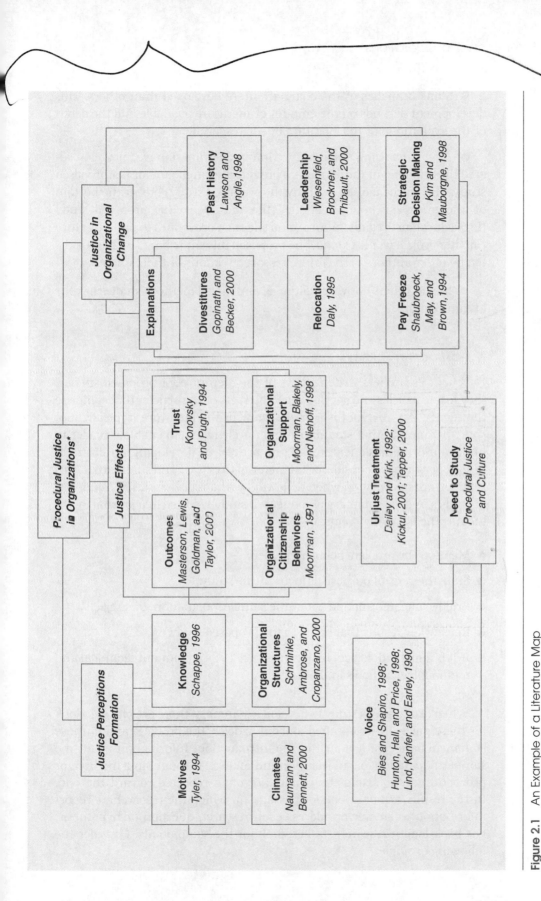

Figure 2.1 An Example of a Literature Map

*Employees' concerns about the fairness of and the making of managerial decisions

SOURCE: Janovec (2001). Reprinted by permission.

● Some branches of the chart are more developed than others. This development depends on the amount of literature available and the depth of the exploration of the literature by the researcher.

● After organizing the literature into a diagram, Janovec next considered the branches of the figure that provide a springboard for her proposed study. She placed a Need to Study (or proposed study) box at the bottom of the map, she briefly identified the nature of this proposed study (Procedural Justice and Culture), and she then drew lines to past literature that her project would extend. She proposed this study based on ideas written by other authors in the future research sections of their studies.

● Include quantitative, qualitative, and mixed methods studies in your literature map.

Abstracting Studies

When researchers write reviews of the literature for proposed studies, they locate articles and develop brief abstracts of the articles that comprise the review. An **abstract** is a brief review of the literature (typically in a short paragraph) that summarizes major elements, to enable a reader to understand the basic features of the article. In developing an abstract, researchers need to consider what material to extract and summarize. This is important information when reviewing perhaps dozens, if not hundreds, of studies. A good summary of a research study reported in a journal might include the following points:

● Mention the problem being addressed.

● State the central purpose or focus of the study.

● Briefly state information about the sample, population, or subjects.

● Review key results that relate to the proposed study.

● If it is a methodological review (Cooper, 1984), point out technical and methodological flaws in the study.

When examining a study to develop a summary, there are places to look for these parts. In well-crafted journal articles, the problem and purpose statements are clearly stated in the introduction. Information about the sample, population, or subjects is found midway through, in a method (or procedure) section, and the results are often reported toward the end. In the results sections, look for passages in which the researchers report information to answer or address each research question or hypothesis. For book-length research studies, look for the same points. Consider the following example:

Example 2.1   *Literature Review in a Quantitative Study*

Here follows a paragraph summarizing the major components of a quantitative study (Creswell, Seagren, & Henry, 1979), much like the paragraph might appear in a review of the literature section of a dissertation or a journal article. In this passage, I have chosen key components to be abstracted.

Creswell, Seagren, and Henry (1979) tested the Biglan model, a three-dimensional model clustering 36 academic areas into hard or soft, pure or applied, life or nonlife areas, as a predictor of chairpersons' professional development needs. Eighty department chairpersons located in four state colleges and one university of a Midwestern state participated in the study. Results showed that chairpersons in different academic areas differed in terms of their professional development needs. Based on the findings, the authors recommended that those who develop inservice programs need to consider differences among disciplines when they plan for programs.

My colleagues and I began with an in-text reference in accord with the format in the APA (2001) style manual. Next, we reviewed the central purpose of the study, followed by information about the data collection. The abstract ended by stating the major results and presenting the practical implications of these results.

How are essays, opinions, typologies, and syntheses of past research abstracted, since these are not research studies? The material to be extracted from these nonempirical studies would be as follows:

● Mention the problem being addressed by the article or book.

● Identify the central theme of the study.

● State the major conclusions related to this theme.

● If the review type is methodological, mention flaws in reasoning, logic, force of argument, and so forth.

Consider the following example that illustrates the inclusion of these aspects:

Example 2.2   *Literature Review in a Study Advancing a Typology*

Sudduth (1992) completed a quantitative dissertation in political science on the topic of the use of strategic adaptation in rural hospitals. He reviewed

(Continued)

(Continued)

the literature in several chapters at the beginning of the study. In an example of summarizing a single study advancing a typology, Sudduth summarized the problem, the theme, and the typology:

> Ginter, Duncan, Richardson, and Swayne (1991) recognize the impact of the external environment on a hospital's ability to adapt to change. They advocate a process that they call environmental analysis which allows the organization to strategically determine the best responses to change occurring in the environment. However, after examining the multiple techniques used for environmental analysis, it appears that no comprehensive conceptual scheme or computer model has been developed to provide a complete analysis of environmental issues (Ginter et al., 1991). The result is an essential part of strategic change that relies heavily on a non-quantifiable and judgmental process of evaluation. To assist the hospital manager to carefully assess the external environment, Ginter et al. (1991) have developed the typology given in Figure 2.1.
>
> (Sudduth, 1992, p. 44)

In this example, the authors referenced the study with an in-text reference, mentioned the problem ("a hospital's ability to adapt to change"), identified the central theme ("a process that they call environmental analysis"), and stated the conclusions related to this theme (e.g., "no comprehensive conceptual model," "developed the typology").

Style Manuals

In both examples, I have introduced the idea of using appropriate APA style for referencing the article at the beginning of the abstract. **Style manuals** provide guidelines for creating a scholarly style of a manuscript, such as a consistent format for citing references, creating headings, presenting tables and figures, and using nondiscriminatory language. A basic tenet in reviewing the literature is to use an appropriate and consistent reference style throughout. When identifying a useful document, make a complete reference to the source using an appropriate style. For dissertation proposals, graduate students should seek guidance from faculty, dissertation committee members, or department or college officials about the appropriate style manual to use for citing references.

The *Publication Manual of the American Psychological Association, Fifth Edition* (APA, 2001) is the most popular style manual used in the fields of education and psychology. The University of Chicago (*A Manual of Style*, 1982), Turabian (1973), and Campbell and Ballou (1977) are also used, but less widely than the APA style in the social sciences. Some journals have developed their own variations of the popular styles. I recommend identifying a style that is acceptable for your writing audiences and adopting it early in the planning process.

The most important style considerations involve in-text, end-of-text, heading, and figures and tables use. Some suggestions for using style manuals for scholarly writing are these:

● When writing *in-text* references, keep in mind the appropriate form for types of references and pay close attention to the format for multiple citations.

● When writing the *end-of-text* references, note whether the style manual calls for them to be alphabetized or numbered. Also, cross-check that each in-text reference is included in the end-of-text list.

● The *headings* are ordered in a scholarly paper in terms of levels. First, note how many levels of headings you will have in your research study. Then, refer to the style manual for the appropriate format for each. Typically, research reports contains between two and four levels of headings.

● If *footnotes* are used, consult the style manual for their proper placement. Footnotes are used less frequently in scholarly papers today than a few years ago. If you include them, note whether they go at the bottom of the page, the end of each chapter, or at the end of the paper.

● *Tables* and *figures* have a specific form in each style manual. Note such aspects as bold lines, titles, and spacing in the examples given.

In summary, the most important aspect of using a style manual is to be consistent in the approach throughout the manuscript.

The Definition of Terms

Another topic related to reviewing the literature is the identification and definition of terms that readers will need in order to understand a proposed research project. A **definition of terms** section may be found separate from the literature review, included as part of the literature review, or placed in different sections of a proposal.

Define terms that individuals outside the field of study may not understand and that go beyond common language (Locke, Spirduso, & Silverman, 2007). Clearly, whether a term should be defined is a matter of judgment, but define a term if there is any likelihood that readers will not know its meaning. Also, define terms when they first appear so that a reader does not read ahead in the proposal operating with one set of definitions only to find out later that the author is using a different set. As Wilkinson (1991) commented, "scientists have sharply defined terms with which to think clearly about their research and to communicate their findings and ideas accurately" (p. 22). Defining terms also adds precision to a scientific study, as Firestone (1987) states:

> The words of everyday language are rich in multiple meanings. Like other symbols, their power comes from the combination of meaning in a specific setting. . . . Scientific language ostensibly strips this multiplicity of meaning from words in the interest of precision. This is the reason common terms are given "technical meanings" for scientific purposes. (p. 17)

With this need for precision, one finds terms stated early in the introduction to articles. In dissertations and thesis proposals, terms are typically defined in a special section of the study. The rationale is that in formal research, students must be precise in how they use language and terms. The need to ground thoughts in authoritative definitions constitutes good science.

Define terms introduced in all sections of the research plan:

- The title of the study

- The problem statement

- The purpose statement

- The research questions, hypotheses, or objectives

- The literature review

- The theory base of the study

- The methods section

Special terms that need to be defined appear in all three types of studies: qualitative, quantitative, and mixed methods.

- In *qualitative* studies, because of the inductive, evolving methodological design, inquirers may define few terms at the beginning, though may advance tentative definitions. Instead, themes (or perspectives or dimensions) may emerge through the data analysis. In the procedure section, authors define these terms in the procedure section as they surface during the process of research. This approach is to delay the definition of terms until they appear in the study and it makes such definitions difficult to specify in research proposals. For this reason, qualitative proposals often do not include separate sections for definition of terms, but the writers pose tentative, qualitative definitions before their entry into the field.

● On the other hand, *quantitative* studies—operating more within the deductive model of fixed and set research objectives—include extensive definitions early in the research proposal. Investigators place them in separate sections and precisely define them. The researchers try to comprehensively define all relevant terms at the beginning of studies and to use accepted definitions found in the literature.

● In *mixed methods studies*, the approach to definitions might include a separate section if the study begins with a first phase of quantitative data collection. If it begins with qualitative data collection, then the terms may emerge during the research, and they are defined in the findings or results section of the final report. If both quantitative and qualitative data collection occurs at the same time, then the priority given to one or the other will govern the approach for definitions. However, in all mixed methods studies, there are terms that may be unfamiliar to readers—for example, the definition of a mixed methods study itself, in a procedural discussion (see Chapter 10). Also, clarify terms related to the strategy of inquiry used, such as concurrent or sequential, and the specific name for it (e.g., concurrent triangulation design, as discussed in Chapter 10).

No one approach governs how one defines the terms in a study, but several suggestions follow (see also Locke et al., 2007):

● Define a term when it first appears in the proposal. In the introduction, for example, a term may require definition to help the reader understand the research problem and questions or hypotheses in the study.

● Write definitions at a specific operational or applied level. Operational definitions are written in specific language rather than abstract, conceptual definitions. Since the definition section in a dissertation provides an opportunity for the author to be specific about the terms used in the study, a preference exists for operational definitions.

● Do not define the terms in everyday language; instead, use accepted language available in the research literature. In this way, the terms are grounded in the literature and not invented (Locke et al., 2007). It is possible that the precise definition of a term is not available in the literature and everyday language will need to be used. In this case, provide a definition and use the term consistently throughout the plan and the study (Wilkinson, 1991).

● Researchers might define terms so that they accomplish different goals. A definition may describe a common language word (e.g., organization). It may also be paired with a limitation, such as, "The *curriculum* will be limited to those after school activities that the current *School District Manual* lists as approved for secondary school students" (Locke et al.,

2007, p. 130). It may establish a criterion that will be used in the study, such as, "*High grade point average* means a cumulative GPA of 3.7 or above on a 4.0 scale." It could also define a term operationally, such as, "*Reinforcement* will refer to the procedure of listing all club members in the school newspaper, providing special hall passes for members, and listing club memberships on school transcripts" (Locke et al., 2007, p. 130).

● Although no one format exists for defining terms, one approach is to develop a separate section, called the "Definition of Terms," and clearly set off the terms and their definitions by highlighting the term. In this way, the word is assigned an invariant meaning (Locke et al., 2007). Typically, this separate section is not more than two to three pages in length.

Two examples illustrate varied structures for defining terms in a research study:

Example 2.3   *Terms Defined in a Mixed Methods Dissertation*

This first example illustrates a lengthy definition of terms presented in a mixed methods study in a separate section of Chapter 1, which introduces the study. VanHorn-Grassmeyer (1998) studied how 119 new professionals in student affairs in colleges and universities engage in reflection, either individually or collaboratively. She both surveyed the new professionals and conducted in-depth interviews with them. Because she studied individual and collaborative reflection among student affairs professionals, she provided detailed definitions of these terms in the beginning of the study. I illustrate two of her terms below. Notice how she referenced her definitions in meanings formed by other authors in the literature:

Individual Reflection

Schon (1983) devoted an entire book to concepts he named reflective thinking, reflection-in-action, and reflective practice; this after an entire book was written a decade earlier with Argyris (Argyris & Schon, 1978) to introduce the concepts. Therefore, a concise definition of this researcher's understanding of individual reflection that did justice to something that most aptly had been identified as an intuitive act, was difficult to reach. However, the most salient characteristics of individual reflection for the purposes of this study were these three: a) an "artistry of practice" (Schon, 1983), b) how one practices overtly what one knows intuitively, and c) how a professional enhances practice through thoughtful discourse within the mind.

Student Affairs Professional

A professional has been described in many ways. One description identified an individual who exhibited "a high degree of independent judgment, based on a collective, learned body of ideas, perspectives, information, norms, and habits (and who engage(d) in professional knowing)" (Baskett & Marsick, 1992, p. 3). A student affairs professional has exhibited such traits in service to students in a higher education environment, in any one of a number of functions which support academic and co-curricular success.

(VanHorn-Grassmeyer, 1998, pp. 11–12)

Example 2.4   *Terms Defined in an Independent Variables Section*

This second set of two examples illustrates an abbreviated form of writing definitions for a study. The first illustrates a specific operational definition of a key term and the second, the procedural definition of a key term. Vernon (1992) studied how divorce in the middle generation impacts grandparents' relationships with their grandchildren. These definitions were included in a section on independent variables.

Kinship Relationship to the Grandchild

Kinship relationship to the grandchild refers to whether the grandparents are maternal grandparents or paternal grandparents. Previous research (e.g., Cherlin and Furstenberg, 1986) suggests that maternal grandparents tend to be closer to their grandchildren.

Sex of Grandparent

Whether a grandparent is a grand*mother* or grand*father* has been found to be a factor in the grandparent/grandchild relationship (i.e., grandmothers tend to be more involved than grandfathers which is thought to be related to the kinkeeping role of women within the family (e.g., Hagestad, 1988).

(Vernon, 1992, pp. 35–36)

A Quantitative or Mixed Methods Literature Review

When composing a review of the literature, it is difficult to determine how much literature to review. In order to address this problem, I have developed a model that provides parameters around the literature review, especially as it might be designed for a quantitative or mixed methods study that employs a standard literature review section. For a qualitative study, the literature review might explore aspects of the central phenomenon being addressed and divide it into topical areas. But the literature review for a qualitative study, as discussed earlier, can be placed in a proposal in several ways (e.g., as a rationale for the research problem, as a separate section, as something threaded throughout the study, as compared with the results of a project).

For a *quantitative study* or *the quantitative strand of a mixed methods study*, write a review of the literature that contains sections about the literature related to major independent variables, major dependent variables, and studies that relate the independent and dependent variables (more on variables in Chapter 3). This approach seems appropriate for dissertations and for conceptualizing the literature to be introduced in a journal article. Consider a literature review to be composed of five components: an introduction, Topic 1 (about the independent variable), Topic 2 (about the dependent variable), Topic 3, (studies that address both the independent and dependent variables), and a summary. Here is more detail about each section:

1. Introduce the review by telling the reader about the sections included in it. This passage is a statement about the organization of the section.

2. Review Topic 1, which addresses the scholarly literature about the *independent* variable or variables. With several independent variables, consider subsections or focus on the single most important variable. Remember to address only the literature about the independent variable; keep the literature about the independent and dependent variables separate in this model.

3. Review Topic 2, which incorporates the scholarly literature about the *dependent* variable or variables. With multiple dependent variables, write subsections about each variable or focus on a single important dependent variable.

4. Review Topic 3, which includes the scholarly literature that relates the *independent* variable(s) to the *dependent* variable(s). Here we are at the crux of the proposed study. Thus, this section should be relatively short and contain studies that are extremely close in topic to the proposed study. Perhaps nothing has been written on the topic. Construct a section that is as close as possible to the topic or review studies that address the topic at a more general level.

5. Provide a summary that highlights the most important studies, captures major themes, suggests why more research is needed on the topic, and advances how the proposed study will fill this need.

This model focuses the literature review, relates it closely to the variables in the research questions and hypotheses, and sufficiently narrows the study. It becomes a logical point of departure for the research questions and the method section.

SUMMARY

Before searching the literature, identify your topic, using such strategies as drafting a brief title or stating a central research question. Also consider whether this topic can and should be researched by reviewing whether there is access to participants and resources and whether the topic will add to the literature, be of interest to others, and be consistent with personal goals.

Researchers use the scholarly literature in a study to present results of similar studies, to relate the present study to an ongoing dialogue in the literature, and to provide a framework for comparing results of a study with other studies. For qualitative, quantitative, and mixed methods designs, the literature serves different purposes. In qualitative research, the literature helps substantiate the research problem, but it does not constrain the views of participants. A popular approach is to include more literature at the end of a qualitative study than at the beginning. In quantitative research, the literature not only helps to substantiate the problem, but it also suggests possible questions or hypotheses that need to be addressed. A separate literature review section is typically found in quantitative studies. In mixed methods research, the use of literature will depend on the type of design and weight given to the qualitative and quantitative aspects.

When conducting a literature review, identify key words for searching the literature. Then search the online databases, such as ERIC, ProQuest, Google Scholar, PubMed, and more specialized databases, such as PsycINFO, Sociofile, and SSCI. Then, locate articles or books based on a priority of searching first for journal articles and then books. Identify references that will make a contribution to your literature review. Group these studies into a literature map that shows the major categories of studies and positions your proposed study within those categories. Begin writing summaries of the studies, noting complete references according to a style manual (e.g., APA, 2001) and extracting information about the research that includes the research problem, the questions, the data collection and analysis, and the final results.

Define key terms, and possibly develop a definition of terms section for your proposal or include them within your literature review. Finally, consider the overall structure for organizing these studies. One quantitative research model is to divide the review into sections according to major variables (a quantitative approach) or major subthemes of the central phenomenon (a qualitative approach) that you are studying.

Writing Exercises

1. Develop a literature map of the literature on your topic. Include in the map the proposed study and draw lines from the proposed study to other categories of studies so that a reader can easily see how yours will extend existing literature.

2. Organize a review of the literature for a quantitative study and follow the model for delimiting the literature to reflect the variables in the study. As an alternative, organize a review of literature for a qualitative study and include it in an introduction as a rationale for the research problem in the study.

3. Practice using an online computer database to search for the literature on your topic. Conduct several searches until you find an article that is as close as possible to your research topic. Then conduct a second search using descriptors mentioned in this article. Locate 10 articles that you would select and abstract for your literature review.

4. Based on your search results from Exercise 3, write one quantitative and one qualitative abstract of two research studies found in your online search. Use the guidelines provided in this chapter for the elements to include in your abstracts.

ADDITIONAL READINGS

Locke, L. F., Spirduso, W. W., & Silverman, S. J. (2007) *Proposals that work: A guide for planning dissertations and grant proposals* (5th ed.) Thousand Oaks, CA: Sage.

Lawrence Locke, Waneen Spirduso, and Stephen Silverman describe three stages for reviewing the literature: develop the concepts that provide a rationale for the study, develop subtopics for each major concept, and add the most important references that support each concept. They also provide six rules for defining terms in a scholarly study: never invent words, provide definitions early in a proposal, do not use common language forms of words, define words when they are first introduced, and use specific definitions for words.

Merriam, S. B. (1998). *Qualitative research and case study applications in education.* San Francisco: Jossey-Bass.

Sharan Merriam provides an extensive discussion about the use of literature in qualitative studies. She identifies steps in reviewing the literature and poses useful criteria for selecting references. These include checking to see if the author is an authority on the topic, how recently the work was published, whether the resource is relevant to the proposed research topic, and the quality of the resource. Merriam further suggests that the literature review is not a linear process of reading the literature, identifying the theoretical framework, and then writing the problem statement. Instead, the process is highly interactive among these steps.

Punch, K. F. (2005). *Introduction to social research: Quantitative and qualitative approaches* (2nd ed.). London: Sage.

Keith Punch provides a guide to social research that equally addresses quantitative and qualitative approaches. His conceptualizations of central issues that divide the two approaches address key differences. Punch notes that when writing a proposal or report, the point at which to concentrate on the literature varies in different styles of research. Factors that affect that decision include the style of research, the overall research strategy, and how closely the study will follow the directions of the literature.

CHAPTER THREE

The Use of Theory

One component of reviewing the literature is to determine what theories might be used to explore the questions in a scholarly study. In *quantitative research*, researchers often test theories as an explanation for answers to their questions. In a quantitative dissertation, an entire section of a research proposal might be devoted to presenting the theory for the study. In *qualitative research*, the use of theory is much more varied. The inquirer may generate a theory as the final outcome of a study and place it at the end of a project, such as in grounded theory. In other qualitative studies, it comes at the beginning and provides a lens that shapes what is looked at and the questions asked, such as in ethnographies or in advocacy research. In mixed methods research, researchers may both test theories and generate them. Moreover, mixed methods research may contain a theoretical lens, such as a focus on feminist, racial, or class issues, that guides the entire study.

I begin this chapter by focusing on theory use in a quantitative study. It reviews a definition of a theory, the use of variables in a quantitative study, the placement of theory in a quantitative study, and the alternative forms it might assume in a written plan. Procedures in identifying a theory are next presented, followed by a script of a theoretical perspective section of a quantitative research proposal. Then the discussion moves to the use of theory in a qualitative study. Qualitative inquirers use different terms for theories, such as *patterns, theoretical lens*, or *naturalistic generalizations*, to describe the broader explanations used or developed in their studies. Examples in this chapter illustrate the alternatives available to qualitative researchers. Finally, the chapter turns to the use of theories in mixed methods research and the use of a transformative perspective that is popular in this approach.

QUANTITATIVE THEORY USE

Variables in Quantitative Research

Before discussing quantitative theories, it is important to understand variables and the types that are used in forming theories. A **variable** refers to a characteristic or attribute of an individual or an organization that can be measured or observed and that varies among the people or organization being studied (Creswell, 2007a). A variable typically will vary in two or more categories or on a continuum of scores, and it can be measured or assessed on a scale. Psychologists prefer to use the term *construct* (rather than *variable*), which carries the connotation more of an abstract idea than a specifically defined term. However, social scientists typically use the term *variable*, and it will be employed in this discussion. Variables often measured in studies include gender, age, socioeconomic status (SES), and attitudes or behaviors such as racism, social control, political power, or leadership. Several texts provide detailed discussions about the types of variables one can use and their scales of measurement (e.g., Isaac & Michael, 1981; Keppel, 1991; Kerlinger, 1979; Thorndike, 1997). Variables are distinguished by two characteristics: temporal order and their measurement (or observation).

Temporal order means that one variable precedes another in time. Because of this time ordering, it is said that one variable affects or causes another variable, though a more accurate statement would be that one variable *probably* causes another. When dealing with studies in the natural setting and with humans, researchers cannot absolutely prove cause and effect (Rosenthal & Rosnow, 1991), and social scientists now say that there is probable causation. Temporal order means that quantitative researchers think about variables in an order from "left to right" (Punch, 2005) and order the variables in purpose statements, research questions, and visual models into left-to-right, cause-and-effect presentations. Thus,

- *Independent variables* are those that (probably) cause, influence, or affect outcomes. They are also called *treatment, manipulated, antecedent*, or *predictor* variables.

- *Dependent variables* are those that depend on the independent variables; they are the outcomes or results of the influence of the independent variables. Other names for dependent variables are *criterion, outcome*, and *effect* variables.

- *Intervening or mediating variables* stand between the independent and dependent variables, and they mediate the effects of the independent variable on the dependent variable. For example, if students do well on a research methods test (dependent variable), that result may be due to (a) their study preparation (independent variable) and/or (b) their organization of study ideas into a framework (intervening variable) that influenced their performance on the test. The mediating variable, the organization of study, stands between the independent and dependent variables.

- *Moderating variables* are new variables constructed by a researcher by taking one variable and multiplying it by another to determine the joint impact of both (e.g., age X attitudes toward quality of life). These variables are typically found in experiments.

● Two other types of variables are *control variables* and *confounding variables*. Control variables play an active role in quantitative studies. These are a special type of independent variable that researchers measure because they potentially influence the dependent variable. Researchers use statistical procedures (e.g., analysis of covariance) to control for these variables. They may be demographic or personal variables (e.g., age or gender) that need to be "controlled" so that the true influence of the independent variable on the dependent can be determined. Another type of variable, a *confounding (or spurious) variable*, is not actually measured or observed in a study. It exists, but its influence cannot be directly detected. Researchers comment on the influence of confounding variables after the study has been completed, because these variables may have operated to explain the relationship between the independent variable and dependent variable, but they were not or could not be easily assessed (e.g., discriminatory attitudes).

In a quantitative research study, variables are related to answer a research question (e.g., "How does self-esteem influence the formation of friendships among adolescents?") or to make predictions about what the researcher expects the results to show. These predictions are called *hypotheses* (e.g., "Individual positive self-esteem expands the number of friends of adolescents.")

Definition of a Theory

With this background on variables, we can proceed to the use of quantitative theories. In *quantitative* research, some historical precedent exists for viewing a theory as a scientific prediction or explanation (see G. Thomas, 1997, for different ways of conceptualizing theories and how they might constrain thought). For example, Kerlinger's (1979) definition of a theory is still valid today. He said, a theory is "a set of interrelated constructs (variables), definitions, and propositions that presents a systematic view of phenomena by specifying relations among variables, with the purpose of explaining natural phenomena" (p. 64).

In this definition, a **theory** is an interrelated set of constructs (or variables) formed into propositions, or hypotheses, that specify the relationship among variables (typically in terms of magnitude or direction). A theory might appear in a research study as an argument, a discussion, or a rationale, and it helps to explain (or predict) phenomena that occur in the world. Labovitz and Hagedorn (1971) add to this definition the idea of a *theoretical rationale*, which they define as "specifying how and why the variables and relational statements are interrelated" (p. 17). Why would an independent variable, X, influence or affect a dependent variable, Y? The theory would provide the explanation for this expectation or prediction. A discussion about this theory would appear in a section of a proposal on the literature review or on the *theory base, the theoretical rationale,* or *the theoretical perspective.* I prefer the term *theoretical perspective* because it has been popularly used as a required section for proposals for research when one submits an application to present a paper at the American Educational Research Association conference.

The metaphor of a rainbow can help to visualize how a theory operates. Assume that the rainbow *bridges* the independent and dependent variables (or constructs) in a study. This rainbow ties together the variables and provides an overarching explanation for *how* and *why* one would expect the independent variable to explain or predict the dependent variable. Theories develop when researchers test a prediction over and over. For example, here is how the process of developing a theory works. Investigators combine independent, mediating, and dependent variables based on different forms of measures into questions. These questions provide information about the type of relationship (positive, negative, or unknown) and its magnitude (e.g., high or low). Forming this information into a predictive statement (hypothesis), a researcher might write, "The greater the centralization of power in leaders, the greater the disenfranchisement of the followers." When researchers test hypotheses such as this over and over in different settings and with different populations (e.g., the Boy Scouts, a Presbyterian church, the Rotary Club, and a group of high school students), a theory emerges, and someone gives it a name (e.g., a theory of attribution). Thus, theory develops as an explanation to advance knowledge in particular fields (Thomas, 1997).

Another aspect of theories is that they vary in their breadth of coverage. Neuman (2000) reviews theories at three levels: micro-level, meso-level, and macro-level. Micro-level theories provide explanations limited to small slices of time, space, or numbers of people, such as Goffman's theory of face work, which explains how people engage in rituals during face-to-face interactions. Meso-level theories link the micro and macro levels. These are theories of organizations, social movement, or communities, such as Collins's theory of control in organizations. Macro-level theories explain larger aggregates, such as social institutions, cultural systems, and whole societies. Lenski's macro-level theory of social stratification, for example, explains how the amount of surplus a society produces increases with the development of the society.

Theories are found in the social science disciplines of psychology, sociology, anthropology, education, and economics, as well as within many subfields. To locate and read about these theories requires searching literature databases (e.g., *Psychological Abstracts, Sociological Abstracts*) or reviewing guides to the literature about theories (e.g., see Webb, Beals, & White, 1986).

Forms of Theories

Researchers state their theories in research proposals in several ways, such as a series of hypotheses, if–then logic statements, or visual models. First, some researchers state theories in the form of interconnected hypotheses. For example, Hopkins (1964) conveyed his theory of influence processes as a series of 15 hypotheses. Some of the hypotheses are as follows (these have been slightly altered to remove the gender-specific pronouns):

1. The higher one's rank, the greater one's centrality.

2. The greater one's centrality, the greater one's observability.

3. The higher one's rank, the greater one's observability.

4. The greater one's centrality, the greater one's conformity.

5. The higher one's rank, the greater one's conformity.

6. The greater one's observability, the greater one's conformity.

7. The greater one's conformity, the greater one's observability. (p. 51)

A second way is to state a theory as a series of if–then statements that explain why one would expect the independent variables to influence or cause the dependent variables. For example, Homans (1950) explains a theory of interaction:

> If the frequency of interaction between two or more persons increases, the degree of their liking for one another will increase, and vice versa. . . . Persons who feel sentiments of liking for one another will express those sentiments in activities over and above the activities of the external system, and these activities may further strengthen the sentiments of liking. The more frequently persons interact with one another, the more alike in some respects both their activities and their sentiments tend to become. (pp. 112, 118, 120)

Third, an author may present a theory as a visual model. It is useful to translate variables into a visual picture. Blalock (1969, 1985, 1991) advocates causal modeling and recasts verbal theories into causal models so that a reader can visualize the interconnections of variables. Two simplified examples are presented here. As shown in Figure 3.1, three independent variables influence a single dependent variable, mediated by the influence of two intervening variables. A diagram such as this one shows the possible causal sequence among variables leading to modeling through path analysis and more advanced analyses using multiple measures of variables as found in structural equation modeling (see Kline, 1998). At an introductory level, Duncan (1985) provides useful suggestions about the notation for constructing these visual causal diagrams:

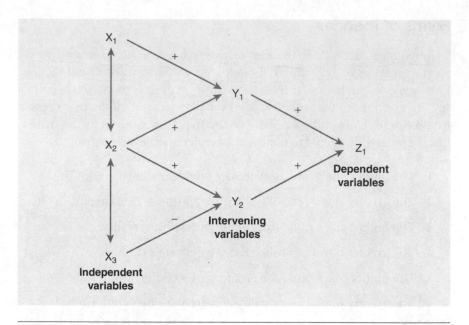

Figure 3.1   Three Independent Variables Influence a Single Dependent
Variable Mediated by Two Intervening Variables

- Position the dependent variables on the right in the diagram and the independent variables on the left.

- Use one-way arrows leading from each determining variable to each variable dependent on it.

- Indicate the strength of the relationship among variables by inserting valence signs on the paths. Use positive or negative valences that postulate or infer relationships.

- Use two-headed arrows connected to show unanalyzed relationships between variables not dependent upon other relationships in the model.

More complicated causal diagrams can be constructed with additional notation. This one portrays a basic model of limited variables, such as typically found in a survey research study.

A variation on this theme is to have independent variables in which control and experimental groups are compared on one independent variable in terms of an outcome (dependent variable). As shown in Figure 3.2, two groups on variable X are compared in terms of their influence on Y, the dependent variable. This design is a between-groups experimental design (see Chapter 8). The same rules of notation previously discussed apply.

These two models are meant only to introduce possibilities for connecting independent and dependent variables to build theories. More complicated designs employ multiple independent and dependent variables in elaborate models of causation (Blalock, 1969, 1985). For example,

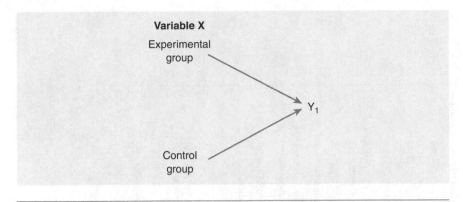

Figure 3.2   Two Groups With Different Treatments on X Are Compared in Terms of Y

Jungnickel (1990), in a doctoral dissertation proposal about research productivity among faculty in pharmacy schools, presented a complex visual model, as shown in Figure 3.3. Jungnickel asked what factors influence a faculty member's scholarly research performance. After identifying these factors in the literature, he adapted a theoretical framework found in nursing research (Megel, Langston, & Creswell, 1988) and developed a visual model portraying the relationship among these factors, following the rules for constructing a model introduced earlier. He listed the independent variables on the far left, the intervening variables in the middle, and the dependent variables on the right. The direction of influence flowed from the left to the right, and he used plus and minus signs to indicate the hypothesized direction.

Placement of Quantitative Theories

In *quantitative* studies, one uses theory deductively and places it toward the beginning of the proposal for a study. With the objective of testing or verifying a theory rather than developing it, the researcher advances a theory, collects data to test it, and reflects on its confirmation or disconfirmation by the results. The theory becomes a framework for the entire study, an organizing model for the research questions or hypotheses and for the data collection procedure. The deductive model of thinking used in a quantitative study is shown in Figure 3.4. The researcher tests or verifies a theory by examining hypotheses or questions derived from it. These hypotheses or questions contain variables (or constructs) that the researcher needs to define. Alternatively, an acceptable definition might be found in the literature. From here, the investigator locates an instrument to use in measuring or observing attitudes or behaviors of participants in a study. Then the investigator collects scores on these instruments to confirm or disconfirm the theory.

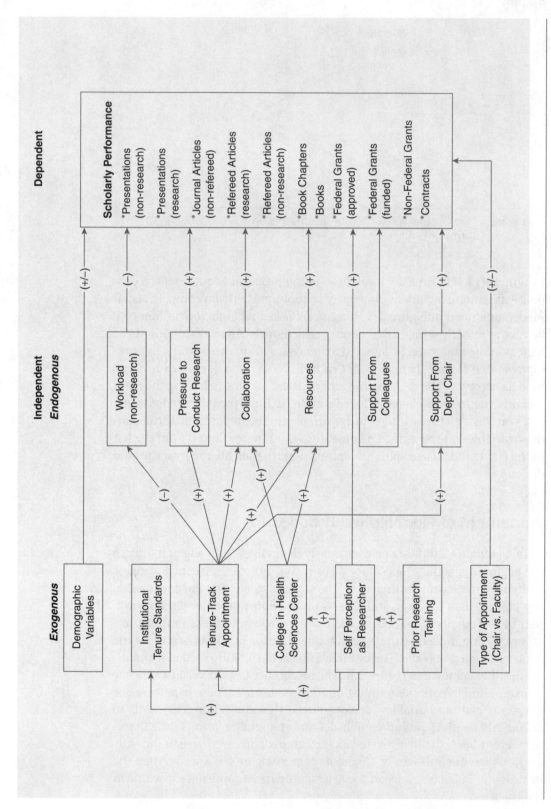

Figure 3.3 A Visual Model of a Theory of Faculty Scholarly Performance

SOURCE: Jungnickel (1990). Reprinted with permission.

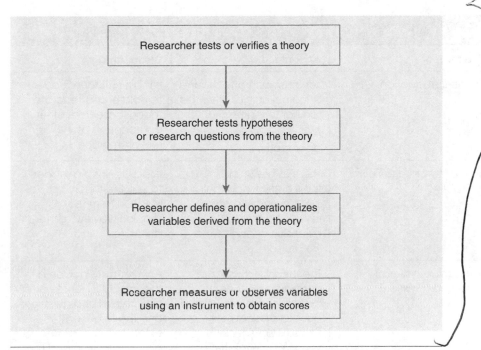

Figure 3.4   The Deductive Approach Typically Used in Quantitative Research

This deductive approach to research in the quantitative approach has implications for the *placement of a theory* in a quantitative research study (see Table 3.1).

A general guide is to introduce the theory early in a plan or study: in the introduction, in the literature review section, immediately after hypotheses or research questions (as a rationale for the connections among the variables), or in a separate section of the study. Each placement has its advantages and disadvantages.

A **research tip**: I write the theory into a separate section in a research proposal so that readers can clearly identify the theory from other components. Such a separate passage provides a complete explication of the theory section, its use, and how it relates to the study.

Writing a Quantitative Theoretical Perspective

Using these ideas, the following presents a model for writing a quantitative theoretical perspective section into a research plan. Assume that the task is to identify a theory that explains the relationship between independent and dependent variables.

1. Look in the discipline-based literature for a theory. If the unit of analysis for variables is an individual, look in the psychology literature; to study groups or organizations, look in the sociological literature. If the

Table 3.1   Options for Placing Theory in a Quantitative Study

Placement	Advantages	Disadvantages
In the introduction	An approach often found in journal articles, it will be familiar to readers. It conveys a deductive approach.	It is difficult for a reader to isolate and separate theory base from other components of the research process.
In the literature review	Theories are found in the literature and their inclusion in a literature review is a logical extension or part of the literature.	It is difficult for a reader to see the theory in isolation from the scholarly review of the literature.
After hypotheses or research questions	The theory discussion is a logical extension of hypotheses or research questions because it explains how and why variables are related.	A writer may include a theoretical rationale after hypotheses and questions and leave out an extended discussion about the origin and use of the theory.
In a separate section	This approach clearly separates the theory from other components of the research process, and it enables a reader to better identify and to understand the theory base for the study.	The theory discussion stands in isolation from other components of the research process and, as such, a reader may not easily connect it with other components of the research process.

project examines individuals and groups, consider the social psychology literature. Of course, theories from other disciplines may be useful, too (e.g., to study an economic issue, the theory may be found in economics).

2. Examine also prior studies that address the topic or a closely related topic. What theories were used by other authors? Limit the number of theories and try to identify *one overarching theory* that explains the central hypothesis or major research question.

3. As mentioned earlier, ask the *rainbow* question that bridges the independent and dependent variables: Why would the independent variable(s) influence the dependent variables?

4. Script out the theory section. Follow these lead sentences: "The theory that I will use is _____ (name the theory). It was developed by _____ (identify the origin, source, or developer of the theory), and it was

used to study _____ (identify the topics where one finds the theory being applied). This theory indicates that _____ (identify the propositions or hypotheses in the theory). As applied to my study, this theory holds that I would expect my independent variable(s) _____ (state independent variables) to influence or explain the dependent variable(s) _____ (state dependent variables) because _____ (provide a rationale based on the logic of the theory)."

Thus, the topics to include in a quantitative theory discussion are the theory to be used, its central hypotheses or propositions, information about past use of the theory and its application, and statements that reflect how it relates to a proposed study. This model is illustrated in the following example by Crutchfield (1986).

Example 3.1   *A Quantitative Theory Section*

Crutchfield (1986) wrote a doctoral dissertation titled *Locus of Control, Interpersonal Trust, and Scholarly Productivity*. Surveying nursing educators, her intent was to determine if locus of control and interpersonal trust affected the levels of publications of the faculty. Her dissertation included a separate section in the introductory chapter titled "Theoretical Perspective," which follows. It includes these points:

- The theory she planned to use

- The central hypotheses of the theory

- Information about who has used the theory and its applicability

- An adaptation of the theory to variables in her study using if–then logic

I have added annotations in italics to mark key passages.

Theoretical Perspective

In formulation of a theoretical perspective for studying the scholarly productivity of faculty, social learning theory provides a useful prototype. This conception of behavior attempts to achieve a balanced synthesis of cognitive psychology with the principles of behavior modification (Bower & Hilgard, 1981). Basically, this unified theoretical framework "approaches the explanation of human behavior in terms of a continuous (reciprocal) interaction between cognitive, behavioral, and environmental determinants" (Bandura, 1977, p. vii). *(Author identifies the theory for the study.)*

(Continued)

(Continued)

While social learning theory accepts the application of reinforcements such as shaping principles, it tends to see the role of rewards as both conveying information about the optimal response and providing incentive motivation for a given act because of the anticipated reward. In addition, the learning principles of this theory place special emphasis on the important roles played by vicarious, symbolic, and self-regulating processes (Bandura, 1971).

Social learning theory not only deals with learning, but seeks to describe how a group of social and personal competencies (so called personality) could evolve out of social conditions within which the learning occurs. It also addresses techniques of personality assessment (Mischel, 1968), and behavior modification in clinical and educational settings (Bandura, 1977; Bower & Hilgard, 1981; Rotter, 1954). *(Author describes social learning theory.)*

Further, the principles of social learning theory have been applied to a wide range of social behavior such as competitiveness, aggressiveness, sex roles, deviance, and pathological behavior (Bandura & Walters, 1963; Bandura, 1977; Mischel, 1968; Miller & Dollard, 1941; Rotter, 1954; Staats, 1975). *(Author describes the use of the theory.)*

Explaining social learning theory, Rotter (1954) indicated that four classes of variables must be considered: behavior, expectancies, reinforcement, and psychological situations. A general formula for behavior was proposed which states: "the potential for a behavior to occur in any specific psychological situation is the function of the expectancy that the behavior will lead to a particular reinforcement in that situation and the value of that reinforcement" (Rotter, 1975, p. 57).

Expectancy within the formula refers to the perceived degree of certainty (or probability) that a causal relationship generally exists between behavior and rewards. This construct of generalized expectancy has been defined as internal locus of control when an individual believes that reinforcements are a function of specific behavior, or as external locus of control when the effects are attributed to luck, fate, or powerful others. The perceptions of causal relationships need not be absolute positions, but rather tend to vary in degree along a continuum depending upon previous experiences and situational complexities (Rotter, 1966). *(Author explains variables in the theory.)*

In the application of social learning theory to this study of scholarly productivity, the four classes of variables identified by Rotter (1954) will be defined in the following manner.

1. Scholarly productivity is the desired behavior or activity.

2. Locus of control is the generalized expectancy that rewards are or are not dependent upon specific behaviors.

3. Reinforcements are the rewards from scholarly work and the value attached to these rewards.

4. The educational institution is the psychological situation which furnishes many of the rewards for scholarly productivity.

With these specific variables, the formula for behavior which was developed by Rotter (1975) would be adapted to read: The potential for scholarly behavior to occur within an educational institution is a function of the expectancy that this activity will lead to specific rewards and of the value that the faculty member places on these rewards. In addition, the interaction of interpersonal trust with locus of control must be considered in relation to the expectancy of attaining rewards through behaviors as recommended in subsequent statements by Rotter (1967). Finally, certain characteristics, such as educational preparation, chronological age, post-doctoral fellowships, tenure, or full-time versus part-time employment may be associated with the scholarly productivity of nurse faculty in a manner similar to that seen within other disciplines. *(Author applied the concepts to her study.)*

The following statement represents the underlying logic for designing and conducting this study. If faculty believe that: (a) their efforts and actions in producing scholarly works will lead to rewards (locus of control), (b) others can be relied upon to follow through on their promises (interpersonal trust), (c) the rewards for scholarly activity are worthwhile (reward values), and (d) the rewards are available within their discipline or institution (institutional setting), then they will attain high levels of scholarly productivity (pp. 12–16). *(Author concluded with the if-then logic to relate the independent variables to the dependent variables.)*

QUALITATIVE THEORY USE

Variation in Theory Use in Qualitative Research

Qualitative inquirers use theory in their studies in several ways. First, much like in quantitative research, it is used as a broad explanation for behavior and attitudes, and it may be complete with variables, constructs, and hypotheses. For example, ethnographers employ cultural themes or "aspects of culture" (Wolcott, 1999, p. 113) to study in their qualitative projects, such as social control, language, stability and change, or social organization, such as kinship or families (see Wolcott's 1999 discussion about texts that address cultural topics in anthropology). Themes in this context provide a ready-made series of hypotheses to be tested from the literature. Although researchers might not refer to them as theories, they provide broad explanations that anthropologists use to study the culture-sharing behavior and attitudes of people. This approach is popular in qualitative health science research in which investigators begin with a theoretical model, such as the adoption of health practices or a quality of life theoretical orientation.

Second, researchers increasingly use a **theoretical lens** or **perspective in qualitative research,** which provides an overall orienting lens for the study of questions of gender, class, and race (or other issues of marginalized groups). This lens becomes an advocacy perspective that shapes the types of questions asked, informs how data are collected and analyzed, and provides a call for action or change. Qualitative research of the 1980s underwent a transformation to broaden its scope of inquiry to include these theoretical lenses. They guide the researchers as to what issues are important to examine (e.g., marginalization, empowerment) and the people that need to be studied (e.g., women, homeless, minority groups). They also indicate how the researcher positions himself or herself in the qualitative study (e.g., up front or biased from personal, cultural, and historical contexts) and how the final written accounts need to be written (e.g., without further marginalizing individuals, by collaborating with participants). In critical ethnography studies, researchers begin with a theory that informs their studies. This causal theory might be one of emancipation or repression (Thomas, 1993).

Some of these qualitative theoretical perspectives available to the researcher are as follows (Creswell, 2007):

● *Feminist perspectives* view as problematic women's diverse situations and the institutions that frame those situations. Research topics may include policy issues related to realizing social justice for women in specific contexts or knowledge about oppressive situations for women (Olesen, 2000).

● *Racialized discourses* raise important questions about the control and production of knowledge, particularly about people and communities of color (Ladson-Billings, 2000).

● *Critical theory* perspectives are concerned with empowering human beings to transcend the constraints placed on them by race, class, and gender (Fay, 1987).

● *Queer theory*—a term used in this literature—focuses on individuals calling themselves lesbians, gays, bisexuals, or transgendered people. The research using this approach does not objectify individuals, is concerned with cultural and political means, and conveys the voices and experiences of individuals who have been suppressed (Gamson, 2000).

● *Disability inquiry* addresses the meaning of inclusion in schools and encompasses administrators, teachers, and parents who have children with disabilities (Mertens, 1998).

Rossman and Rallis (1998) capture the sense of theory as critical and postmodern perspectives in qualitative inquiry:

As the 20th century draws to a close, traditional social science has come under increasing scrutiny and attack as those espousing critical and postmodern perspectives challenge objectivist assumptions and traditional norms for the conduct of research. Central to this attack are four interrelated notions: (a) Research fundamentally involves issues of power; (b) the research report is not transparent but rather it is authored by a raced, gendered, classed, and politically oriented individual; (c) race, class, and gender are crucial for understanding experience; and (d) historic, traditional research has silenced members of oppressed and marginalized groups. (p. 66)

Third, distinct from this theoretical orientation are qualitative studies in which theory (or some other broad explanation) becomes the *end point*. It is an inductive process of building from the data to broad themes to a generalized model or theory (see Punch, 2005). The logic of this inductive approach is shown in Figure 3.5.

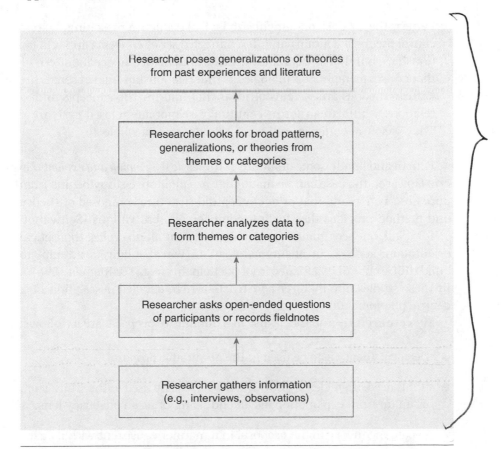

Figure 3.5   The Inductive Logic of Research in a Qualitative Study

The researcher begins by gathering detailed information from participants and then forms this information into categories or themes. These
themes are developed into broad patterns, theories, or generalizations that
are then compared with personal experiences or with existing literature on
the topic.

The development of themes and categories into patterns, theories,
or generalizations suggests varied end points for qualitative studies. For
example, in case study research, Stake (1995) refers to an assertion as a
propositional generalization—the researcher's summary of interpretations
and claims—to which is added the researcher's own personal experiences,
called "naturalistic generalizations" (p. 86). As another example, grounded
theory provides a different end point. Inquirers hope to discover a theory
that is grounded in information from participants (Strauss & Corbin,
1998). Lincoln and Guba (1985) refer to "pattern theories" as explanations that develop during naturalistic or qualitative research. Rather
than the deductive form found in quantitative studies, these pattern theories or generalizations represent interconnected thoughts or parts linked to
a whole.

Neuman (2000) provides additional information about pattern theories:

> Pattern theory does not emphasize logical deductive reasoning. Like
> causal theory, it contains an interconnected set of concepts and rela-
> tionships, but it does not require causal statements. Instead, pattern
> theory uses metaphor or analogies so that relationship "makes sense."
> Pattern theories are systems of ideas that inform. The concepts and
> relations within them form a mutually reinforcing, closed system.
> They specify a sequence of phases or link parts to a whole. (p. 38)

Fourth and finally, some qualitative studies *do not employ any explicit theory.* However, the case can be made that no qualitative study begins from
pure observation and that prior conceptual structure composed of theory
and method provides the starting point for all observations (Schwandt,
1993). Still, one sees qualitative studies that contain no *explicit* theoretical
orientation, such as in phenomenology, in which inquirers attempt to
build the essence of experience from participants (e.g., see Riemen, 1986).
In these studies, the inquirer constructs a rich, detailed description of a
central phenomenon.

My **research tips** on theory use in a qualitative proposal are as follows:

● Decide if theory is to be used in the qualitative proposal.

● If it is used, then identify how the theory will be used in the study, such
as an up-front explanation, as an end point, or as an advocacy lens.

● Locate the theory in the proposal in a manner consistent with its use.

Locating the Theory in Qualitative Research

How theory is used affects its placement in a qualitative study. In those studies with a cultural theme or a theoretical lens, the theory occurs in the opening passages of the study. Consistent with the emerging design of qualitative inquiry, the theory may appear at the beginning and be modified or adjusted based on participant views. Even in the most theory-oriented qualitative design, such as critical ethnography, Lather (1986) qualifies the use of theory:

> Building empirically grounded theory requires a reciprocal relationship between data and theory. Data must be allowed to generate propositions in a dialectical manner that permits use of *a priori* theoretical frameworks, but which keeps a particular framework from becoming the container into which the data must be poured. (p. 267)

Example 3.2   *A Theory Early in a Qualitative Study*

Murguia, Padilla, and Pavel (1991) studied the integration of 24 Hispanic and Native American students into the social system of a college campus. They were curious about how ethnicity influenced social integration, and they began by relating the participants' experiences to a theoretical model, the Tinto model of social integration. They felt that the model had been "incompletely conceptualized and, as a consequence, only imprecisely understood and measured" (p. 133).

Thus, the model was not being tested, as one would find in a quantitative project, but modified. At the end of the study, the authors refined Tinto's model and advanced their modification that described how ethnicity functions. In contrast to this approach, in qualitative studies with an end point of a theory (e.g., a grounded theory), a pattern, or a generalization, the theory emerges at the end of the study. This theory might be presented as a logic diagram, a visual representation of relationships among concepts.

Example 3.3   *A Theory at the End of a Qualitative Study*

Using a national database of 33 interviews with academic department chairpersons, we (Creswell & Brown, 1992) developed a grounded theory interrelating variables (or categories) of chair influence on scholarly performance of faculty. The theory section came into the article as the last section,

(Continued)

(Continued)

where we presented a visual model of the theory developed inductively from categories of information supplied by interviewees. In addition, we also advanced directional hypotheses that logically followed from the model. Moreover, in the section on the model and the hypotheses, we compared the results from participants with results from other studies and the theoretical speculations in the literature. For example, we stated,

This proposition and its sub-propositions represent unusual, even contrary evidence, to our expectations. Contrary to proposition 2.1, we expected that the career stages would be similar not in type of issue but in the range of issues. Instead we found that the issues for post-tenure faculty covered almost all the possible problems on the list. Why were the tenured faculty's needs more extensive than non-tenured faculty? The research productivity literature suggests that one's research performance does not decline with the award of tenure (Holley 1977). Perhaps diffuse career goals of post-tenure faculty expand the possibilities for "types" of issues. In any case, this sub-proposition focuses attention on the understudied career group that Furniss (1981) reminds us needs to be examined in more detail.

(Creswell & Brown, 1992, p. 58)

As this example shows, we developed a visual model that interrelated variables, derived this model inductively from informant comments, and placed the model at the end of the study, where the central propositions in it could be contrasted with the existing theories and literature.

MIXED METHODS THEORY USE

Theory use in mixed methods studies may include theory deductively, in quantitative theory testing and verification, or inductively as in an emerging qualitative theory or pattern. A social science or a health science theory may be used as a framework to be tested in either a quantitative or qualitative approach to inquiry. Another way to think about theory in mixed methods research is as a *theoretical lens* or *perspective* to guide the study. Studies are beginning to emerge that employ mixed methods designs using a lens to study gender, race or ethnicity, disability, sexual orientation, and other bases of diversity (Mertens, 2003).

Historically, the idea of using a theoretical lens in mixed methods research was mentioned by Greene and Caracelli in 1997. They identified the use of a *transformative design* as a distinct form of mixed methods research. This design gave primacy to value-based, action-oriented research, such as in participatory action research and empowerment approaches. In this design, they suggest mixing the value commitments of different traditions (e.g., bias-free from quantitative and bias-laden from qualitative), the use of diverse methods, and a focus on action solutions. The implementation of these ideas in the practice of mixed methods research has now been carried forward by other authors.

More information on procedures has appeared in a chapter written by Creswell, Plano Clark, Gutmann, and Hanson (2003). They identified the use of theoretical perspectives, such as gendered, feminist; cultural/racial/ethnic; lifestyle; critical; and class and social status. These perspectives became an overlay over mixed methods designs (see Chapter 10). They further developed visual models to portray how these lenses might provide a guiding perspective for a mixed methods study. Mertens (2003) continued the discussion. As outlined in Box 3.1, she advocated for the importance of a theory lens in mixed methods research. In detailing a transformative–emancipatory paradigm and specific procedures, she emphasized the role that values played in studying feminist, ethnic/racial, and disability issues. Her transformative theory was an umbrella term for research that was emancipatory, antidiscriminatory, participative, Freirian, feminist, racial/ethnic, for individuals with disabilities, and for all marginalized groups.

Mertens identifies the implications of these transformative theories for mixed methods research. These involve integration of the transformative–emancipatory methodology into all phases of the research process. Reading through the questions in Box 3.1, one gains a sense of the importance of studying issues of discrimination and oppression and of recognizing diversity among study participants. These questions also address treating individuals respectfully through gathering and communicating data collection and through reporting results that lead to changes in social processes and relationships.

Box 3.1   Transformative-Emancipatory Questions for Mixed Methods Researchers Throughout the Research Process

Defining the Problem and Searching the Literature

- Did you deliberately search the literature for concerns of diverse groups and issues of discrimination and oppression?

- Did the problem definition arise from the community of concern?

- Did your mixed methods approach arise from spending quality time with these communities? (i.e., building trust? using an appropriate theoretical framework other than a deficit model? developing balanced—positive and negative—questions? developing questions that lead to transformative answers, such as questions focused on authority and relations of power in institutions and communities?)

(Continued)

(Continued)

Identifying the Research Design

● Does your research design deny treatment to any groups and respect ethical considerations of participants?

Identifying Data Sources and Selecting Participants

● Are the participants of groups associated with discrimination and oppression?

● Are the participants appropriately labeled?

● Is there a recognition of diversity within the target population?

● What can be done to improve the inclusiveness of the sample to increase the probability that traditionally marginalized groups are adequately and accurately represented?

Identifying or Constructing Data Collection Instruments and Methods

● Will the data collection process and outcomes benefit the community being studied?

● Will the research findings be credible to that community?

● Will communication with that community be effective?

● Will the data collection open up avenues for participation in the social change process?

Analyzing, Interpreting, and Reporting and Using Results

● Will the results raise new hypotheses?

● Will the research examine subgroups (i.e., multilevel analyses) to analyze the differential impact on diverse groups?

● Will the results help understand and elucidate power relationships?

● Will the results facilitate social change?

SOURCE: Adapted from D. M. Mertens (2003), "Mixed Methods and the Politics of Human Research: The Transformative-Emancipatory Perspective," in A. Tashakkori & C. Teddlie (Eds.), *Handbook of Mixed Methods in the Social & Behavioral Sciences*. Adapted with permission.

Example 3.4   *Theory in a Transformative–Emancipatory Mixed Methods Study*

Hopson, Lucas, and Peterson (2000) studied issues in an urban, predominantly African American HIV/AIDS community. Consistent with a transformative–emancipatory framework, they examined the language of participants with HIV/AIDS within the participants' social context. They first conducted 75 open-ended ethnographic interviews to identify "language themes" (p. 31), such as blame, ownership, and acceptance or nonacceptance. They also collected 40 semistructured interviews that addressed demographics, daily routine, drug use, knowledge of HIV/AIDS risks, and drug and sexual socio-behavioral characteristics. From this qualitative data, the authors used concepts and questions to refine follow-up questions, including the design of a quantitative postintervention instrument. The authors suggested that empowerment approaches in evaluation can be useful, with researchers listening to the voices of real people and acting on what program participants say.

The design in this study gave "primacy to the value-based and action-oriented dimensions of different inquiry traditions" (Greene & Caracelli, 1997, p. 24) in a mixed methods study. The authors used a theoretical lens for reconfiguring the language and dialogue of participants, and they advanced the importance of empowerment in research.

In using theory in a mixed methods proposal,

- Determine if theory is to be used.

- Identify its use in accord with quantitative or qualitative approaches.

- If theory is used as in a transformational strategy of inquiry, define this strategy and discuss the points in the proposed study in which the emancipatory ideas will be used.

SUMMARY

Theory has a place in quantitative, qualitative, and mixed methods research. Researchers use theory in a quantitative study to provide an explanation or prediction about the relationship among variables in the study. Thus, it is essential to have grounding in the nature and use of variables as they form research questions and hypotheses. A theory explains how and why the variables are related, acting as a bridge between or among the variables. Theory may be broad or narrow in scope, and researchers state their theories in several ways, such as a series of hypotheses, if–then logic statements, or visual models. Using theories deductively, investigators advance them at the beginning of the study in the literature review. They also include them with the hypotheses or research questions or place them in a separate section. A script can help design the theory section for a research proposal.

In qualitative research, inquirers employ theory as a broad explanation, much like in quantitative research, such as in ethnographies. It may also be a theoretical lens or perspective that raises questions related to gender, class, race, or some combination of these. Theory also appears as an end point of a qualitative study, a generated theory, a pattern, or a generalization that emerges inductively from data collection and analysis. Grounded theorists, for example, generate a theory grounded in the views of participants and place it as the conclusion of their studies. Some qualitative studies do not include an explicit theory and present descriptive research of the central phenomenon.

Mixed methods researchers use theory either deductively (as in quantitative research) or inductively (as in qualitative research). Writers also are beginning to identify the use of theoretical lenses or perspectives (e.g., related to gender, lifestyle, race/ethnicity, and class) in their mixed methods studies. A transformational–emancipatory design incorporates this perspective, and recent developments have identified procedures for incorporating this perspective into all phases of the research process.

Writing Exercises

1. Write a theoretical perspective section for your research plan following the script for a quantitative theory discussion presented in this chapter.

2. For a quantitative proposal you are planning, draw a visual model of the variables in the theory using the procedures for causal model design advanced in this chapter.

3. Locate qualitative journal articles that (a) use an a priori theory that is modified during the process of research, (b) generate or develop a theory at the end of the study, and (c) represent descriptive research without the use of an explicit theoretical model.

4. Locate a mixed methods study that uses a theoretical lens, such as a feminist, ethnic/racial, or class perspective. Identify specifically how the lens shapes the steps taken in the research process, using Box 3.1 as a guide.

WRITING EXERCISES

ADDITIONAL READINGS

Flinders, D. J., & Mills, G. E. (Eds.). (1993). *Theory and concepts in qualitative research: Perspectives from the field.* New York: Teachers College Press, Teachers College, Columbia University.

David Flinders and Geoffrey Mills have edited a book about perspectives from the field—"theory at work"—as described by different qualitative researchers. The chapters illustrate little consensus about defining theory and whether it is a vice or virtue. Further, theory operates at many levels in research, such as formal theories, epistemological theories, methodological theories, and meta-theories. Given this diversity, it is best to see actual theory at work in qualitative studies, and this volume illustrates practice from critical, personal, formal, and educational criticism.

Mertens, D. M. (2003). Mixed methods and the politics of human research: The transformative-emancipatory perspective. In A. Tashakkori & C. Teddlie (Eds.), *Handbook of mixed methods in social & behavioral research* (pp. 135–164). Thousand Oaks, CA: Sage.

Donna Mertens recognizes that historically, research methods have not concerned themselves with the issues of the politics of human research and social justice. Her chapter explores the transformative–emancipatory paradigm of research as a framework or lens for mixed methods research as it has emerged from scholars from diverse ethnic/racial groups, people with disabilities, and feminists. A unique aspect of her chapter is how she weaves together this paradigm of thinking and the steps in the process of conducting mixed methods research.

Thomas, G. (1997). What's the use of theory? *Harvard Educational Review, 67*(1), 75–104.

Gary Thomas presents a reasoned critique of the use of theory in educational inquiry. He notes the various definitions of theory and maps out four broad uses of theory: (a) as thinking and reflection, (b) as tighter or looser hypotheses, (c) as explanations for adding to knowledge in different fields, and (d) as formally expressed statements in science. Having noted these uses, he then embraces the thesis that theory unnecessarily structures and constrains thought. Instead, ideas should be in a constant flux and should be "ad hocery," as characterized by Toffler.

CHAPTER FOUR

Writing Strategies and Ethical Considerations

Before designing a proposal, it is important to have an idea of the general structure or outline of the topics and their order. The structure will differ depending on whether you write a quantitative, qualitative, or mixed methods project. Another general consideration is to be aware of good writing practices that will help to ensure a consistent and highly readable proposal (and research project). Throughout the project, it is important to engage in ethical practices and to anticipate what ethical issues will likely arise. This chapter provides outlines for the overall structure of proposals, writing practices that make them easy to read, and ethical issues that need to be considered as the proposals are written.

WRITING THE PROPOSAL

Sections in a Proposal

It is helpful to consider the topics that will go into a proposal. All the topics need to be interrelated and provide a cohesive picture of the entire project. An outline is helpful, but the topics will differ depending on whether the proposal is for a qualitative, quantitative, or mixed methods study. In this chapter, I advance outlines for sections of a proposal, as an overview of the process. In individual chapters to follow, the sections will be detailed further.

Overall, however, there are central arguments that frame any proposal. They are introduced as nine central arguments by Maxwell (2005). I pose them here as questions to be addressed in a scholarly proposal.

1. What do readers need, to better understand your topic?

2. What do readers know little about in terms of your topic?

3. What do you propose to study?

4. What is the setting and who are the people that you will study?

5. What methods do you plan to use to provide data?

6. How will you analyze the data?

7. How will you validate your findings?

8. What ethical issues will your study present?

9. What do preliminary results show about the practicability and value of the proposed study?

These nine questions, if adequately addressed in one section for each question, constitute the foundation of good research, and they could provide the overall structure for a proposal. The inclusion of validating findings, ethical considerations (to be addressed shortly), the need for preliminary results, and early evidence of practical significance focus a reader's attention on key elements often overlooked in discussions about proposed projects.

Format for a Qualitative Proposal

In the light of these points, I propose two alternative models. Example 4.1 is drawn from a constructivist/interpretivist perspective, whereas Example 4.2 is based more on an advocacy/participatory model of qualitative research.

Example 4.1   *A Qualitative Constructivist/Interpretivist Format*

Introduction

Statement of the problem (including existing literature about the problem, significance of the study)

Purpose of the study and how study will be delimited

The research questions

Procedures

Philosophical assumptions of qualitative research

Qualitative research strategy

Role of the researcher

Data collection procedures

Strategies for validating findings

Proposed narrative structure of the study

Anticipated ethical issues

Preliminary pilot findings (if available)

Expected outcomes

Appendixes: Interview questions, observational forms, timeline, and proposed budget

In this example, the writer includes only two major sections, the introduction and the procedures. A review of the literature may be included, but it is optional, and, as discussed in Chapter 3, the literature may be included to a greater extent at the end of the study or in the expected outcomes section. I have added sections that may at first seem unusual. Developing a timeline for the study and presenting a proposed budget provide useful information to graduate committees, although these sections are typically not found in outlines for proposals.

Example 4.2   *A Qualitative Advocacy/Participatory Format*

Introduction

Statement of the problem (including the advocacy/participatory issue being addressed, existing literature about the problem, significance of study)

Purpose of the study and delimitations of the study

The research questions

Procedures

Philosophical assumptions of qualitative research

Qualitative research strategy

Role of the researcher

Data collection procedures (including the collaborative approaches used with participants)

Data recording procedures

(Continued)

(Continued)

    Data analysis procedures

    Strategies for validating findings

    Narrative structure

Anticipated ethical issues

Significance of the study

Preliminary pilot findings (if available)

Expected advocacy/participatory changes

Appendixes: Interview questions, observational forms, timeline, and proposed budget

This format is similar to the constructivist/interpretivist format except that the inquirer identifies a specific advocacy/participatory issue being explored in the study (e.g., marginalization, empowerment), advances a collaborative form of data collection, and mentions the anticipated changes that the research study will likely bring.

Format for a Quantitative Proposal

For a quantitative study, the format conforms to sections typically found in quantitative studies reported in journal articles. The form generally follows the model of an introduction, a literature review, methods, results, and discussion. In planning a quantitative study and designing a dissertation proposal, consider the following format to sketch the overall plan (see Example 4.3).

Example 4.3   *A Quantitative Format*

Introduction

    Statement of the problem (issue, significance of issue)

    Purpose of the study and delimitations

    Theoretical perspective

    Research questions or hypotheses

Review of the literature

Methods

    Type of research design

    Population, sample, and participants

    Data collection instruments, variables, and materials

    Data analysis procedures

Anticipated ethical issues in the study

Preliminary studies or pilot tests

Appendixes: Instruments, timeline, and proposed budget

Example 4.3 is a standard format for a social science study, although the order of the sections, especially in the introduction, may vary from study to study (see, for example, Miller, 1991; Rudestam & Newton, 2007). It presents a useful model for designing the sections for a plan for a dissertation or sketching the topics for a scholarly study.

Format for a Mixed Methods Proposal

In a mixed methods design format, the researcher brings together approaches that are included in both the quantitative and qualitative formats (see Creswell & Plano Clark, 2007). An example of such a format appears in Example 4.4 (adapted from Creswell & Plano Clark, 2007).

Example 4.4   *A Mixed Methods Format*

Introduction

    The research problem

    Past research on the problem

    Deficiencies in past research and one deficiency related to the need to collect both quantitative and qualitative data

    The audiences that will profit from the study

Purpose

    The purpose or study aim of the project and reasons for a mixed methods study

(Continued)

(Continued)

   The research questions and hypotheses (quantitative questions or hypotheses, qualitative questions, mixed methods questions)

   Philosophical foundations for using mixed methods research

   Literature review (review quantitative, qualitative, and mixed methods studies)

Methods

   A definition of mixed methods research

   The type of design used and its definition

   Challenges in using this design and how they will be addressed

   Examples of use of the type of design

   Reference and inclusion of a visual diagram

   Quantitative data collection and analysis

   Qualitative data collection and analysis

   Mixed methods data analysis procedures

   Validity approaches in both quantitative and qualitative research

Researcher's resources and skills

Potential ethical issues

Timeline for completing the study

References and appendixes with instruments, protocols, visuals

---

This format shows that the researcher poses both a purpose statement and research questions for quantitative and qualitative components, as well as mixed components. It is important to specify early in the proposal the reasons for the mixed methods approach and to identify key elements of the design, such as the type of mixed methods study, a visual picture of the procedures, and both the quantitative and qualitative data collection procedures and analysis.

Designing the Sections of a Proposal

Here are several **research tips** that I give to students about designing the overall structure of a proposal.

● Specify the sections early in the design of a proposal. Work on one section will often prompt ideas for other sections. First develop an outline and then write something for each section rapidly, to get ideas down on paper. Then refine the sections as you consider in more detail the information that should go into each one.

● Find proposals that other students have authored under your adviser and look at them closely. Ask your adviser for copies of proposals that he or she especially liked and felt were scholarly products to take to committees. Study the topics addressed and their order as well as the level of detail used in composing the proposal.

● Determine whether your program or institution offers a course on proposal development or some similar topic. Often such a class will be helpful as a support system for your project as well as providing individuals that can react to your proposal ideas as they develop.

● Sit down with your adviser and go over his or her preferred format for a proposal. The order of sections found in published journal articles may not provide the information desired by your adviser or graduate committee.

WRITING IDEAS

Over the years, I have collected books on how to write, and I typically have a new one that I am reading as I work on my research projects. As I work on this third edition, I am reading *Reading Like a Writer* by Francine Prose (Prose, 2006). By reading books such as this, I am constantly reminded of good writing principles that need to go into my research writing. My books span a wide spectrum, from professional trade books to academic writing books. In this section, I have extracted the key ideas that have been meaningful to me from many favorite writing books that I have used.

Writing as Thinking

One sign of inexperienced writers is that they prefer to discuss their proposed study rather than write about it. I recommend the following:

● *Early in the process of research, write ideas down rather than talk about them.* Writing specialists see writing as thinking (Bailey, 1984). Zinsser (1983) discusses the need to get words out of our heads and onto paper. Advisers react better when they read the ideas on paper than when they hear and discuss a research topic with a student or colleague. When a researcher renders ideas on paper, a reader can visualize the final product, actually see how it looks, and begin to clarify ideas. The concept of working ideas out on paper has served many experienced writers well. Before designing a proposal, draft a one- to-two-page overview of your project and have your adviser approve the direction of your proposed study. This draft might contain the essential information: the research problem being addressed, the purpose of the study, the central questions being asked, the source of data, and the significance of the project for different audiences. It might also be useful to draft several one- to-two-page statements on different topics and see which one your adviser likes best and feels would make the best contribution to your field.

● *Work through several drafts of a proposal rather than trying to polish the first draft.* It is illuminating to see how people think on paper. Zinsser (1983) identified two types of writers: the "bricklayer," who makes every paragraph just right before going on to the next paragraph, and the "let-it-all-hang-out-on-the-first-draft" writer, who writes an entire first draft not caring how sloppy it looks or how badly it is written. In between would be someone like Peter Elbow (Elbow, 1973), who recommends that one should go through the iterative process of writing, reviewing, and rewriting. He cites this exercise: With only 1 hour to write a passage, write four drafts (one every 15 minutes) rather than one draft (typically in the last 15 minutes) during the hour. Most experienced researchers write the first draft carefully but do not work for a polished draft; the polish comes relatively late in the writing process.

● *Do not edit your proposal at the early-draft stage.* Instead, consider Franklin's (1986) three-stage model, which I have found useful in developing proposals and in my scholarly writing:

1. First, develop an outline—it could be a sentence or word outline or a visual map.

2. Write out a draft and then shift and sort ideas, moving around entire paragraphs in the manuscript.

3. Finally, edit and polish each sentence.

The Habit of Writing

Establish the discipline or **habit of writing** in a regular and continuous way on your proposal. Although setting aside a completed draft of the proposal for a time may provide some perspective to review your work before final polishing, a start-and-stop process of writing often disrupts the flow of work. It may turn a well-meaning researcher into what I call a weekend writer, an individual who only has time for working on research on weekends after all the "important" work of the week has been accomplished. Continual work on the proposal is writing something each day or at least being engaged daily in the processes of thinking, collecting information, and reviewing that goes into manuscript and proposal production.

Select a time of day to work that is best for you, then use discipline to write at this time each day. Choose a place free of distractions. Boice (1990, pp. 77–78) offers ideas about establishing good writing habits:

- With the aid of the priority principle, make writing a daily activity, regardless of mood, regardless of readiness to write.

- If you feel you do not have time for regular writing, begin by charting your daily activities for a week or two in half-hour blocks. It's likely you'll find a time to write.

- Write while you are fresh.

- Avoid writing in binges.

- Write in small, regular amounts.

- Schedule writing tasks so that you plan to work on specific, manageable units of writing in each session.

- Keep daily charts. Graph at least three things: (a) time spent writing, (b) page equivalents finished, and (c) percentage of planned task completed.

- Plan beyond daily goals.

- Share your writing with supportive, constructive friends until you feel ready to go public.

- Try to work on two or three writing projects concurrently so that you do not become overloaded with any one project.

It is also important to acknowledge that writing moves along slowly and that a writer must ease into writing. Like the runner who stretches before a road race, the writer needs warm-up exercises for both the mind and the fingers. Some leisurely writing activity, such as writing a letter to a friend, brainstorming on the computer, reading some good writing, or studying a favorite poem, can make the actual task of writing easier. I am reminded of John Steinbeck's (1969) "warm-up period" (p. 42) described in detail in *Journal of a Novel: The East of Eden Letters.* Steinbeck began each writing day by writing a letter to his editor and close friend, Pascal Covici, in a large notebook supplied by Covici.

Other exercises may prove useful as warm-ups. Carroll (1990) provides examples of exercises to improve a writer's control over descriptive and emotive passages:

- Describe an object by its parts and dimensions, without first telling the reader its name.

- Write a conversation between two people on any dramatic or intriguing subject.

- Write a set of directions for a complicated task.

- Take a subject and write about it three different ways (pp. 113–116).

This last exercise seems appropriate for qualitative researchers who analyze their data for multiple codes and themes (see Chapter 9 for qualitative data analysis).

Consider also the writing implements and the physical location that aid the process of disciplined writing. The implements—a computer, a yellow legal-sized pad, a favorite pen, a pencil, even coffee and Triscuits (Wolcott, 2001)—offer the writer options for ways to be comfortable when writing. The physical setting can also help. Annie Dillard, the Pulitzer prize-winning novelist, avoided appealing workplaces:

> One wants a room with no view, so imagination can meet memory in the dark. When I furnished this study seven years ago, I pushed the long desk against a blank wall, so I could not see from either window. Once, fifteen years ago, I wrote in a cinder-block cell over a parking lot. It overlooked a tar-and-gravel roof. This pine shed under trees is not quite so good as the cinder-block study was, but it will do. (Dillard, 1989, pp. 26–27)

Readability of the Manuscript

Before beginning the writing of a proposal, consider how you will enhance the readability of it for other people. The APA (2001) *Publication Manual* discusses an orderly presentation by showing the relationships between ideas and through the use of transitional words. In addition, it is important to use consistent terms, a staging and foreshadowing of ideas, and coherence built into the plan.

- Use *consistent terms* throughout the proposal. Use the same term each time a variable is mentioned in a quantitative study or a central phenomenon in a qualitative study. Refrain from using synonyms for these terms, a problem that causes the reader to work at understanding the meaning of ideas and to monitor subtle shifts in meaning.

- Consider how narrative thoughts of different types guide a reader. This concept was advanced by Tarshis (1982), who recommended that writers stage thoughts to guide readers. These were of four types:

1. Umbrella thoughts—the general or core ideas one is trying to get across

2. Big thoughts in writing—specific ideas or images that fall within the realm of umbrella thoughts and serve to reinforce, clarify, or elaborate upon the umbrella thoughts

3. Little thoughts—ideas or images whose chief function is to reinforce big thoughts

4. Attention or interest thoughts—ideas whose purposes are to keep the reader on track, organize ideas, and keep an individual's attention

Beginning researchers seem to struggle most with umbrella and attention thoughts. A proposal may include too many umbrella ideas, with the content not sufficiently detailed to support large ideas. This might occur in a literature review in which the researcher needs to provide fewer small sections and more larger sections that tie together large bodies of literature. A clear mark of this problem is a continual shift of ideas from one major idea to another in a manuscript. Often, one will see short paragraphs in introductions to proposals, like those written by journalists in newspaper articles. Thinking in terms of a detailed narrative to support umbrella ideas may help this problem.

Attention thoughts, those that provide organizational statements to guide the reader, are also needed. Readers need road signs to guide them from one major idea to the next (Chapters 6 and 7 of this book discuss major road signs in research, such as purpose statements and research questions and hypotheses). An organizing paragraph is often useful at the beginning and end of literature reviews. Readers need to see the overall organization of the ideas through introductory paragraphs and to be told the most salient points they should remember in a summary.

● Use *coherence* to add to the readability of the manuscript. **Coherence in writing** means that the ideas tie together and logically flow from one sentence to another and from one paragraph to another. For example, the repetition of the same variable names in the title, the purpose statement, the research questions, and the review of the literature headings in a quantitative project illustrates this thinking. This approach builds coherence into the study. Emphasizing a consistent order whenever independent and dependent variables are mentioned also reinforces this idea.

On a more detailed level, coherence builds through connecting sentences and paragraphs in the manuscript. Zinsser (1983) suggests that every sentence should be a logical sequel to the one that preceded it. The hook-and-eye exercise (Wilkinson, 1991) is useful for connecting thoughts from sentence to sentence and paragraph to paragraph.

The following passage from a draft of a student's proposal shows a high level of coherence. It comes from the introductory section of a qualitative dissertation project about at-risk students. In this passage, I have taken the liberty of drawing hooks and eyes to connect the ideas from sentence to sentence and from paragraph to paragraph. As mentioned, the objective of the hook-and-eye exercise (Wilkinson, 1991) is to connect major thoughts of each sentence and paragraph. If such a connection cannot easily be made, the written passage lacks coherence, the ideas and topics shift, and the writer needs to add transitional words, phrases, or sentences to establish a clear connection.

In my proposal development classes, I provide a passage from an introduction to a proposal and ask students to connect the sentences using circles for key ideas and lines to connect these key ideas from sentence to sentence. It is important for a reader to find coherence in a proposal starting with the first page. I first give my students an unmarked passage and then, after the exercise, provide a marked passage. Since the key idea of one sentence should connect to a key idea in the next sentence, they need to mark this relationship in the passage. If the sentences do not connect, then transition words are missing that need to be inserted. I also ask students to make sure that the paragraphs are connected with hooks and eyes as well as individual sentences.

Example 4.5   *An Illustration of the Hook-and-Eye Technique*

They sit in the back of the room not because they want to but because it was the place designated to them. Invisible barriers that exist in most classrooms divide the room and separate the students. At the front of the room are the "good" students, who wait with their hands poised ready to fly into the air at a moment's notice. Slouched down like giant insects caught in educational traps, the athletes and their following occupy the center of the room. Those less sure of themselves and their position within the room sit in the back and around the edge of the student body.

The students seated in the outer circle make up a population whom for a variety of reasons are not succeeding in the American public education system. They have always been part of the student population. In the past they have been called disadvantaged, low achieving, retards, impoverished, laggards and a variety of other titles (Cuban, 1989; Presseisen, 1988). Today they are called students-at-risk. Their faces are changing and in urban settings their numbers are growing (Hodgkinson, 1985).

In the past eight years there has been an unprecedented amount of research on the need for excellence in education and the at-risk student. In 1983 the government released a document entitled A Nation At-Risk that identified problems within the American education system and called for major reform. Much of the early reform focused on more vigorous courses of study and higher standards of student achievement (Barber, 1987). In the midst of attention to excellence, it became apparent the needs of the marginal student were not being met. The question of what it would take to guarantee that all students have a fair chance at a quality education was receiving little attention (Hamilton, 1987; Toch, 1984). As the push for excellence in education increased, the needs of the at-risk student became more apparent.

Much of the early research focused on identifying characteristics of the at-risk student (OERI, 1987; Barber & McClellan, 1987; Hahn, 1987; Rumberger, 1987), while others in educational research called for reform and developed programs for at-risk students (Mann, 1987; Presseisen, 1988; Whelage, 1988; Whelege & Lipman, 1988; Stocklinski, 1991; and Levin, 1991). Studies and research on this topic have included experts within the field of education, business and industry as well as many government agencies.

Although progress has been made in identifying characteristics of the at-risk students and in developing programs to meet their needs, the essence of the at-risk issue continues to plague the American school system. Some educators feel that we do not need further research (DeBlois, 1989; Hahn, 1987). Others call for a stronger network between business and education (DeBlois, 1989; Mann, 1987; Whelege, 1988). Still others call for total restructuring of our education system (OERI, 1987; Gainer, 1987; Levin, 1988; McCune, 1988).

After all the research and studies by the experts, we still have students hanging on to the fringe of education. The uniqueness of this study will shift the focus from causes and curriculum to the student. It is time to question the students and to listen to their responses. This added dimension should bring further understanding to research already available and lead to further areas of reform. Dropouts and potential dropouts will be interviewed in depth to discover if there are common factors within the public school setting that interfere with their learning process. This information should be helpful to both the researcher who will continue to look for new approaches in education and the practitioner who works with these students every day.

Voice, Tense, and "Fat"

From working with broad thoughts and paragraphs, I move on to the level of writing sentences and words. Similar grammar and sentence construction issues are addressed in the APA (2001) *Publication Manual*, but I include this section to highlight some common grammar issues that I have seen in student proposals and in my own writing.

My thoughts are directed toward the polish level of writing, to use Franklin's (1986) term. It is a stage addressed late in the writing process. One can find an abundance of writing books about research writing and literary writing with rules and principles to follow concerning good sentence construction and word choice. Wolcott (2001), a qualitative ethnographer, for example, talks about honing editorial skills to eliminate unnecessary words, deleting the passive voice, scaling down qualifiers, eliminating overused phrases, and reducing excessive quotations, use of italics, and parenthetical comments. The following additional ideas about active voice, verb tense, and reduced "fat" can strengthen and invigorate scholarly writing for dissertation and thesis proposals.

● Use the active voice as much as possible in scholarly writing (APA, 2001). According to the literary writer, Ross-Larson (1982), "If the subject acts, the voice is active. If the subject is acted on, the voice is passive" (p. 29). In addition, a sign of passive construction is some variation of an auxiliary verb, such as *was*. Examples include *will be, have been,* and *is being.* Writers can use the passive construction when the person acting can logically be left out of the sentence and when what is acted on is the subject of the rest of the paragraph (Ross-Larson, 1982).

● Use strong verbs and verb tenses appropriate for the passage. Lazy verbs are those that lack action (*is* or *was*, for example) or those used as adjectives or adverbs.

● A common practice exists in using the past tense to review the literature and report results of a study. The future tense would be appropriate at all other times in research proposals and plans. For completed studies, use the present tense to add vigor to a study, especially in the introduction. The APA (2001) *Publication Manual* recommends the past tense (e.g., "Jones reported") or the present perfect tense (e.g., "Researchers have reported") for the literature review and procedures based on past events, the past tense to describe results (e.g., "stress lowered self-esteem"), and the present tense (e.g., "the qualitative findings show") to discuss the results and to present the conclusions. I see this not as a hard and fast rule but as a useful guideline.

● Expect to edit and revise drafts of a manuscript to trim the fat. "**Fat**" means additional words that are unnecessary to convey the meaning of ideas. Writing multiple drafts of a manuscript is standard practice for most writers. The process typically consists of writing, reviewing, and editing. In the editing process, trim excess words from sentences, such as piled-up modifiers, excessive prepositions, and the-of constructions—for example, "the study of"—that add unnecessary verbiage (Ross-Larson, 1982). I was reminded of the unnecessary prose that comes into writing by the example mentioned by Bunge (1985):

> Nowadays you can almost see bright people struggling to reinvent the complex sentence before your eyes. A friend of mine who is a college administrator every now and then has to say a complex sentence, and he will get into one of those morasses that begins, "I would hope that we would be able . . . " He never talked that way when I first met him, but even at his age, at his distance from the crisis in the lives of younger people, he's been to some extent alienated from easy speech. (Bunge, 1985, p. 172)

Begin studying good writing about research using qualitative, quantitative, and mixed methods designs. In good writing, the eye does not pause and the mind does not stumble on a passage. In this present book, I have attempted to draw examples of good prose from human and social science journals, such as *American Journal of Sociology, Journal of Applied Psychology, Administrative Science Quarterly, American Educational Research Journal, Sociology of Education,* and *Image: Journal of Nursing Scholarship.* In the qualitative area, good literature serves to illustrate clear prose and detailed passages. Individuals who teach qualitative research assign well-known books from literature, such as *Moby Dick, The Scarlet Letter,* and *The Bonfire of the Vanities,* as reading assignments (Webb & Glesne, 1992). *Qualitative Inquiry, Qualitative Research, Symbolic Interaction, Qualitative Family Research,* and *Journal of Contemporary Ethnography* represent good, scholarly journals to examine. When using mixed methods research, examine journals that report studies with combined qualitative and quantitative research and data, including many social science journals, such as *Journal of Mixed Methods Research, Field Methods,* and *Quality and Quantity.* Examine the numerous articles cited in the *Handbook of Mixed Methods in the Social & Behavioral Sciences* (Tashakkori & Teddlie, 2003).

ETHICAL ISSUES TO ANTICIPATE

In addition to conceptualizing the writing process for a proposal, researchers need to anticipate the ethical issues that may arise during their studies (Hesse-Bieber & Leavey, 2006). Research does involve collecting data from people, about people (Punch, 2005). As mentioned earlier, writing about these issues is required in making an argument for a study as well as being an important topic in the format for proposals. Researchers need to protect their research participants; develop a trust with them; promote the integrity of research; guard against misconduct and impropriety that might reflect on their organizations or institutions; and cope with new, challenging problems (Isreal & Hay, 2006). Ethical questions are apparent today in such issues as personal disclosure, authenticity and credibility of the research report, the role of researchers in cross-cultural contexts, and issues of personal privacy through forms of Internet data collection (Isreal & Hay, 2006).

In the literature, ethical issues arise in discussions about codes of professional conduct for researchers and in commentaries about ethical dilemmas and their potential solutions (Punch, 2005). Many national associations have published standards or **codes of ethics** on their Web sites for professionals in their fields. For example, see

● Ethical Principles of Psychologists and Code of Conduct, written in 2002, available at www.apa.org/ethics

- The American Sociological Association Code of Ethics, adopted in 1997, available at www.asanet.org

- The American Anthropological Association's Code of Ethics, approved in June 1998, available at www.aaanet.org

- The American Educational Research Association Ethical Standards of the American Educational Research Association, 2002, available at www.aera.net

- The American Nurses Association Code of Ethics for Nurses–Provisions, approved in June 2001, and available at www.ana.org

Ethical practices involve much more than merely following a set of static guidelines, such as those provided by professional associations. Writers need to anticipate and address any ethical dilemmas that may arise in their research (e.g., see Berg, 2001; Punch, 2005; and Sieber, 1998). These issues apply to qualitative, quantitative, and mixed methods research and to all stages of research. Proposal writers need to anticipate them and actively address them in their research plans. In the chapters that follow in Part II, I refer to ethical issues in many stages of research. By mentioning them at this point, I hope to encourage the proposal writer to actively design them into sections of a proposal. Although these discussions will not comprehensively cover all ethical issues, they address major ones. These issues arise primarily in specifying the research problem (Chapter 5); identifying a purpose statement and research questions (Chapters 6 and 7); and collecting, analyzing, and writing up the results of data (Chapters 8, 9, and 10).

Ethical Issues in the Research Problem

Hesse-Biber and Leavy (2006) ask, "How do ethical issues enter into your selection of a research problem?" (p. 86). In writing an introduction to a study, the researcher identifies a significant problem or issue to study and presents a rationale for its importance. During the identification of the research problem, it is important to identify a problem that will benefit individuals being studied, one that will be meaningful for others besides the researcher (Punch, 2005). A core idea of action/participatory research is that the inquirer will not further marginalize or disempower the study participants. To guard against this, proposal developers can conduct pilot projects to establish trust and respect with the participants so that inquirers can detect any marginalization before the proposal is developed and the study begun.

Ethical Issues in the Purpose and Questions

In developing the purpose statement or the central intent and questions for a study, proposal developers need to convey the purpose of the study that will be described to the participants (Sarantakos, 2005). Deception occurs when participants understand one purpose but the researcher has a different purpose in mind. It is also important for researchers to specify the sponsorship of their study. For example, in designing cover letters for survey research, sponsorship is an important element in establishing trust and credibility for a mailed survey instrument.

Ethical Issues in Data Collection

As researchers anticipate data collection, they need to respect the participants and the sites for research. Many ethical issues arise during this stage of the research.

Do not put participants at risk, and respect vulnerable populations. Researchers need to have their research plans reviewed by the Institutional Review Board (IRB) on their college and university campuses. IRB committees exist on campuses because of federal regulations that provide protection against human rights violations. For a researcher, the IRB process requires assessing the potential for risk, such as physical, psychological, social, economic, or legal harm (Sieber, 1998), to participants in a study. Also, the researcher needs to consider the special needs of vulnerable populations, such as minors (under the age of 19), mentally incompetent participants, victims, persons with neurological impairments, pregnant women or fetuses, prisoners, and individuals with AIDS. Investigators file research proposals containing the procedures and information about the participants with the IRB campus committee so that the board can review the extent to which the research being proposed subjects individuals to risk. In addition to this proposal, the researcher develops an **informed consent form** for participants to sign before they engage in the research. This form acknowledges that participants' rights will be protected during data collection. Elements of this consent form include the following (Sarantakos, 2005):

- Identification of the researcher

- Identification of the sponsoring institution

- Indication of how the participants were selected

- Identification of the purpose of the research

- Identification of the benefits for participating

- Identification of the level and type of participant involvement

- Notation of risks to the participant

- Guarantee of confidentiality to the participant

- Assurance that the participant can withdraw at any time

- Provision of names of persons to contact if questions arise

One issue to anticipate about confidentiality is that some participants may not want to have their identity remain confidential. By permitting this, the researcher allows the participants to retain ownership of their voices and exert their independence in making decisions. They do, however, need to be well informed about the possible risks of nonconfidentiality, such as the inclusion of data in the final report that they may not have expected, information that infringes on the rights of others that should remain concealed, and so forth (Giordano, O'Reilly, Taylor, & Dogra, 2007).

● Other ethical procedures during data collection involve gaining the agreement of individuals in authority (e.g., gatekeepers) to provide access to study participants at research sites. This often involves writing a letter that identifies the extent of time, the potential impact, and the outcomes of the research. Use of Internet responses gained through electronic interviews or surveys needs permission from participants. This might be gained through first obtaining permission and then sending out the interview or survey.

● Researchers need to respect research sites so that they are left undisturbed after a research study. This requires that inquirers, especially in qualitative studies involving prolonged observation or interviewing at a site, be cognizant of their impact and minimize their disruption of the physical setting. For example, they might time visits so that they intrude little on the flow of activities of participants. Also, organizations often have guidelines that provide guidance for conducting research without disturbing their settings.

● In experimental studies, investigators need to collect data so that all participants, not only an experimental group, benefit from the treatments. This may require providing *some* treatment to all groups or staging the treatment so that ultimately all groups receive the beneficial treatment.

● An ethical issue arises when there is not reciprocity between the researcher and the participants. Both the researcher and the participants should benefit from the research. In some situations, power can easily be abused and participants can be coerced into a project. Involving individuals collaboratively in the research may provide reciprocity. Highly collaborative studies, popular in qualitative research, may engage participants as co-researchers throughout the research process, such as the design, data collection and analysis, report writing, and dissemination of the findings (Patton, 2002).

● Interviewing in qualitative research is increasingly being seen as a moral inquiry (Kvale, 2007). As such, interviewers need to consider how the interview will improve the human situation (as well as enhance scientific knowledge), how a sensitive interview interaction may be stressful for the participants, whether participants have a say in how their statements are interpreted, how critically the interviewees might be questioned, and what the consequences of the interview for the interviewees and the groups to which they belong might be.

● Researchers also need to anticipate the possibility of harmful, intimate information being disclosed during the data collection process. It is difficult to anticipate and try to plan for the impact of this information during or after an interview (Patton, 2002). For example, a student may discuss parental abuse or prisoners may talk about an escape. Typically in these situations, the ethical code for researchers (which may be different for schools and prisons) is to protect the privacy of the participants and to convey this protection to all individuals involved in a study.

Ethical Issues in Data Analysis and Interpretation

When the researcher analyzes and interprets both quantitative and qualitative data, issues emerge that call for good ethical decisions. In anticipating a research study, consider the following:

● How will the study protect the anonymity of individuals, roles, and incidents in the project? For example, in survey research, investigators disassociate names from responses during the coding and recording process. In qualitative research, inquirers use aliases or pseudonyms for individuals and places, to protect identities.

● Data, once analyzed, need to be kept for a reasonable period of time (e.g., Sieber, 1998, recommends 5–10 years). Investigators should then discard the data so that it does not fall into the hands of other researchers who might misappropriate it.

● The question of who owns the data once it is collected and analyzed also can be an issue that splits research teams and divides individuals against each other. A proposal might mention this issue of ownership and discuss how it will be resolved, such as through the development of a clear understanding between the researcher, the participants, and possibly the faculty advisers (Punch, 2005). Berg (2001) recommends the use of personal agreements to designate ownership of research data. An extension of this idea is to guard against sharing the data with individuals not involved in the project.

● In the interpretation of data, researchers need to provide an accurate account of the information. This accuracy may require debriefing between the researcher and participants in quantitative research (Berg, 2001). It may include, in qualitative research, using one or more of the strategies to check the accuracy of the data with participants or across different data sources (see validation strategies in Chapter 9).

Ethical Issues in Writing and Disseminating the Research

The ethical issues do not stop with data collection and analysis; issues apply as well to the actual writing and dissemination of the final research report. For example,

● Discuss how the research will not use language or words that are biased against persons because of gender, sexual orientation, racial or ethnic group, disability, or age. The APA (2001) *Publication Manual* suggests three guidelines. First, present unbiased language at an appropriate level of specificity (e.g., rather than say, "The client's behavior was typically male," state, "the client's behavior was _____ [specify]"). Second, use language that is sensitive to labels (e.g., rather than "400 Hispanics", indicate "400 Mexicans, Spaniards, and Puerto Ricans"). Third, acknowledge participants in a study (e.g., rather than "subject," use the word "participant," and rather than "woman doctor" use "doctor" or "physician").

● Other ethical issues in writing the research will involve the potential of suppressing, falsifying, or inventing findings to meet a researcher's or an audience's needs. These fraudulent practices are not accepted in professional research communities, and they constitute scientific misconduct (Neuman, 2000). A proposal might contain a proactive stance by the researcher to not engage in these practices.

● In planning a study, it is important to anticipate the repercussions of conducting the research on certain audiences and not to misuse the results to the advantage of one group or another. The researcher needs to provide those at the research site with a preliminary copy of any publications from the research (Creswell, 2007).

● An important issue in writing a scholarly manuscript is to not exploit the labor of colleagues and to provide authorship to individuals who substantially contribute to publications. Isreal and Hay (2006) discuss the unethical practice of so-called gift authorship to individuals who do not contribute to a manuscript and ghost authorship, in which junior staff who made significant contributions have been omitted from the list of authors.

● Finally, it is important to release the details of the research with the study design so that readers can determine for themselves the credibility of the study (Neuman, 2000). Detailed procedures for quantitative, qualitative, and mixed methods research will be emphasized in the chapters to follow. Also, researchers should not engage in duplicate or redundant publication in which authors publish papers that present exactly the same data, discussions, and conclusions and do not offer new material. Some biomedical journals now require authors to declare whether they have published or are preparing to publish papers that are closely related to the manuscript that has been submitted (Isreal & Hay, 2006).

SUMMARY

It is helpful to consider how to write a research proposal before actually engaging in the process. Consider the nine arguments advanced by Maxwell (2005) as the key elements to include and then use one of the four topical outlines provided to craft a thorough qualitative, quantitative, or mixed methods proposal.

In proposal development, begin putting words down on paper to think through ideas; establish the habit of writing on a regular basis; and use strategies such as applying consistent terms, different levels of narrative thoughts, and coherence to strengthen writing. Writing in the active voice, using strong verbs, and revising and editing will help as well.

Before writing the proposal, it is useful to consider the ethical issues that can be anticipated and described in the proposal. These issues relate to all phases of the research process. With consideration for participants, research sites, and potential readers, studies can be designed that contain ethical practices.

Writing Exercises

1. Develop a topical outline for a quantitative, qualitative, or mixed methods proposal. Include the major topics in the examples included in this chapter.

2. Locate a journal article that reports qualitative, quantitative, or mixed methods research. Examine the introduction to the article and, using the hook-and-eye method illustrated in this chapter, identify the flow of ideas from sentence to sentence and from paragraph to paragraph and any deficiencies.

3. Consider one of the following ethical dilemmas that may face a researcher. Describe ways you might anticipate the problem and actively address it in your research proposal.

   a. A prisoner you are interviewing tells you about a potential breakout at the prison that night. What do you do?

   b. A researcher on your team copies sentences from another study and incorporates them into the final written report for your project. What do you do?

   c. A student collects data for a project from several individuals interviewed in families in your city. After the fourth interview, the student tells you that approval has not been received for the project from the Institutional Review Board. What do you do?

WRITING EXERCISES

ADDITIONAL READINGS

Maxwell, J. (2005). *Qualitative research design: An interactive approach.* (2nd ed.). Thousand Oaks, CA: Sage.

Joe Maxwell provides a good overview of the proposal development process for qualitative research that is applicable in many ways to quantitative and mixed methods research as well. He states that a proposal is an argument to conduct a study and presents an example that describes nine necessary steps. Moreover, he includes a complete qualitative proposal and analyzes it as an illustration of a good model to follow.

Sieber, J. E. (1998). Planning ethically responsible research. In L. Bickman & D. J. Rog (Eds.), *Handbook of applied social research methods* (pp. 127–156). Thousand Oaks, CA: Sage.

Joan Sieber discusses the importance of ethical planning as integral to the process of research design. In this chapter, she provides a comprehensive review of many topics related to ethical issues, such as IRBs, informed consent, privacy, confidentiality, and anonymity, as well as elements of research risk and vulnerable populations. Her coverage is extensive, and her recommendations for strategies are numerous.

Isreal, M., & Hay, I. (2006). *Research ethics for social scientists: Between ethical conduct and regulatory compliance.* London: Sage.

Mark Isreal and Lain Hay provide a thoughtful analysis of the practical value of thinking seriously and systematically about what constitutes ethical conduct in the social sciences. They review the different theories of ethics, such as the consequentialist and the nonconsequentialist approaches, virtue ethics, and normative and care-oriented approaches to ethical conduct. They also offer an international perspective, drawing on the history of ethical practices in countries around the world. Throughout the book, they offer practical case examples and ways researchers might treat the cases ethically. In the appendix, they provide three case examples and then call upon leading scholars to comment about how they would approach the ethical issue.

Wolcott, H. F. (2001). *Writing up qualitative research* (2nd ed.). Thousand Oaks, CA: Sage.

Harry Wolcott, an educational ethnographer, has compiled an excellent resource guide addressing numerous aspects of the writing process in qualitative research. It surveys techniques useful in getting started in writing; developing details; linking with the literature, theory, and method; tightening up with revising and editing; and finishing the process by attending to such aspects as the title and appendixes. For all aspiring writers, this is an essential book, regardless of whether a study is qualitative, quantitative, or mixed methods.

PART II

Designing Research

- **Chapter 5**
  The Introduction

- **Chapter 6**
  The Purpose Statement

- **Chapter 7**
  Research Questions and Hypotheses

- **Chapter 8**
  Quantitative Methods

- **Chapter 9**
  Qualitative Procedures

- **Chapter 10**
  Mixed Methods Procedures

This section relates the three designs—quantitative, qualitative, and mixed methods—to the steps in the process of research. Each chapter addresses a separate step in this process.

CHAPTER FIVE

The Introduction

After having decided on a qualitative, quantitative, or mixed methods approach and after conducting a preliminary literature review and deciding on a format for a proposal, the next step in the process is to design or plan the study. A process of organizing and writing out ideas begins, starting with designing an introduction to a proposal. This chapter discusses the composition and writing of a scholarly introduction and examines the differences in writing an introduction for these three different types of designs. Then the discussion turns to the five components of writing a good introduction: (a) establishing the problem leading to the study, (b) reviewing the literature about the problem, (c) identifying deficiencies in the literature about the problem, (d) targeting an audience and noting the significance of the problem for this audience, and (e) identifying the purpose of the proposed study. These components comprise a *social science deficiency model* of writing an introduction, because a major component of the introduction is to set forth the deficiencies in past research. To illustrate this model, a complete introduction in a published research study is presented and analyzed.

THE IMPORTANCE OF INTRODUCTIONS

An introduction is the first passage in a journal article, dissertation, or scholarly research study. It sets the stage for the entire study. As Wilkinson (1991) mentions,

> The introduction is the part of the paper that provides readers with the background information for the research reported in the paper. Its purpose is to establish a framework for the research, so that readers can understand how it is related to other research. (p. 96)

The introduction establishes the issue or concern leading to the research by conveying information about a problem. Because it is the initial passage in a study or proposal, special care must be given to writing it. The introduction needs to create reader interest in the topic, establish the problem that leads to the study, place the study within the larger context of the scholarly literature, and reach out to a specific audience. All of this is achieved in a concise section of a few pages. Because of the messages they must convey and the limited space allowed, introductions are challenging to write and understand.

A **research problem** is the problem or issue that leads to the need for a study. It can originate from many potential sources. It might spring from an experience researchers have had in their personal lives or workplaces. It may come from an extensive debate that has appeared in the literature. It might develop from policy debates in government or among top executives. The sources of research problems are often multiple. Identifying and stating the research problem that underlies a study is not easy: For example, to identify the issue of teenage pregnancy is to point to a problem for women and for society at large. Unfortunately, too many authors of do not clearly identify the research problem, leaving the reader to decide the importance of the issue. When the problem is not clear, it is difficult to understand the significance of the research. Furthermore, the research problem is often confused with the research questions—those questions that the investigator would like answered in order to understand or explain the problem.

To this complexity is added the need for introductions to carry the weight of encouraging the reader to read farther and to see significance in the study.

Fortunately, there is a model for writing a good, scholarly social science introduction. Before introducing this model, it is necessary to distinguish subtle differences between introductions for qualitative, quantitative, and mixed methods studies.

QUALITATIVE, QUANTITATIVE, AND MIXED METHODS INTRODUCTIONS

A general review of all introductions shows that they follow a similar pattern: The authors announce a problem and they justify why it needs to be studied. The type of problem presented in an introduction will vary depending on the approach (see Chapter 1). In a *qualitative* project, the author will describe a research problem that can best be understood by exploring a concept or phenomenon. I have suggested that qualitative research is exploratory, and researchers use it to explore a topic when the variables and theory base are unknown. For example, Morse (1991) says this:

Characteristics of a qualitative research problem are: (a) the concept is "immature" due to a conspicuous lack of theory and previous research; (b) a notion that the available theory may be inaccurate, inappropriate, incorrect, or biased; (c) a need exists to explore and describe the phenomena and to develop theory; or (d) the nature of the phenomenon may not be suited to quantitative measures. (p. 120)

For example, urban sprawl (a problem) needs to be explored because it has not been examined in certain areas of a state. Alternatively, kids in elementary classrooms have anxiety that interferes with learning (a problem), and the best way to explore this problem is to go to schools and visit directly with teachers and students. Some qualitative researchers have a theoretical lens through which the problem will be examined (e.g., the inequality of pay among women and men or the racial attitudes involved in profiling drivers on the highways). Thomas (1993) suggests that "critical researchers begin from the premise that all cultural life is in constant tension between control and resistance" (p. 9). This theoretical orientation shapes the structure of an introduction. Beisel (1990), for example, proposed to examine how the theory of class politics explained the lack of success of an anti-vice campaign in one of three American cities. Thus, within some qualitative studies, the approach in the introduction may be less inductive while still relying on the perspective of participants, like most qualitative studies. In addition, qualitative introductions may begin with a personal statement of experiences from the author, such as those found in phenomenological studies (Moustakas, 1994). They also may be written from a personal, first-person, subjective point of view in which the researcher positions herself or himself in the narrative.

Less variation is seen in *quantitative* introductions. In a quantitative project, the problem is best addressed by understanding what factors or variables influence an outcome. For example, in response to worker cutbacks (a problem for all employees), an investigator may seek to discover what factors influence businesses to downsize. Another researcher may need to understand the high divorce rate among married couples (a problem) and examine whether financial issues contribute to divorce. In both of these situations, the research problem is one in which understanding the factors that explain or relate to an outcome helps the investigator best understand and explain the problem. In addition, in quantitative introductions, researchers sometimes advance a theory to test, and they will incorporate substantial reviews of the literature to identify research questions that need to be answered. A quantitative introduction may be written from the impersonal point of view and in the past tense, to convey objectivity.

A *mixed methods* study can employ either the qualitative or the quantitative approach (or some combination) to writing an introduction. In any given mixed methods study, the emphasis might tip in the direction of either quantitative or qualitative research, and the introduction will mirror that emphasis. For other mixed methods projects, the emphasis will be equal between qualitative and quantitative research. In this case, the problem may be one in which a need exists to both understand the relationship among variables in a situation and explore the topic in further depth. A mixed methods project may initially seek to explain the relationship between smoking behavior and depression among adolescents, then explore the detailed views of adolescents and display different patterns of smoking and depression. With the first phase of this project as quantitative, the introduction may emphasize a quantitative approach with inclusion of a theory that predicts this relationship and a substantive review of the literature.

A MODEL FOR AN INTRODUCTION

These differences among the various approaches are small, and they relate largely to the different types of problems addressed in qualitative, quantitative, and mixed methods studies. It should be helpful to illustrate an approach to designing and writing an introduction to a research study that researchers might use regardless of their approach.

The **deficiencies model of an introduction** is a general template for writing a good introduction. It is a popular approach used in the social sciences, and once its structure is elucidated, the reader will find it appearing repeatedly in many published research studies. It consists of five parts, and a separate paragraph can be devoted to each part, for an introduction of about two pages in length:

1. The research problem

2. Studies that have addressed the problem

3. Deficiencies in the studies

4. The significance of the study for particular audiences

5. The purpose statement

An Illustration

Before a review of each part, here is an excellent example from a quantitative study published by Terenzini, Cabrera, Colbeck, Bjorklund, and Parente (2001) in *The Journal of Higher Education* and titled "Racial and Ethnic Diversity in the Classroom" (reprinted with permission). Following each major section in the introduction, I briefly highlight the component being addressed.

Since passage of the Civil Rights Act of 1964 and the Higher Education Act of 1965, America's colleges and universities have struggled to increase the racial and ethnic diversity of their students and faculty members, and "affirmative action" has become the policy-of-choice to achieve that heterogeneity. *(Authors state the narrative hook.)* These policies, however, are now at the center of an intense national debate. The current legal foundation for affirmative action policies rests on the 1978 *Regents of the University of California v. Bakke* case, in which Justice William Powell argued that race could be considered among the factors on which admissions decisions were based. More recently, however, the U.S. Court of Appeals for the Fifth Circuit, in the 1996 *Hopwood v. State of Texas* case, found Powell's argument wanting. Court decisions turning affirmative action policies aside have been accompanied by state referenda, legislation, and related actions banning or sharply reducing race-sensitive admissions or hiring in California, Florida, Louisiana, Maine, Massachusetts, Michigan, Mississippi, New Hampshire, Rhode Island, and Puerto Rico (Healy, 1998a, 1998b, 1999).

In response, educators and others have advanced educational arguments supporting affirmative action, claiming that a diverse student body is more educationally effective than a more homogeneous one. Harvard University President Neil Rudenstine claims that the "fundamental rationale for student diversity in higher education (is) its educational value" (Rudenstine, 1999, p. 1). Lee Bollinger, Rudenstine's counterpart at the University of Michigan, has asserted, "A classroom that does not have a significant representation from members of different races produces an impoverished discussion" (Schmidt, 1998, p. A32). These two presidents are not alone in their beliefs. A statement published by the Association of American Universities and endorsed by the presidents of 62 research universities stated: "We speak first and foremost as educators. We believe that our students benefit significantly from education that takes place within a diverse setting" ("On the Importance of Diversity in University Admissions," *The New York Times,* April 24, 1997, p. A27). *(Authors identify the research problem.)*

Studies of the impact of diversity on student educational outcomes tend to approach the ways students encounter "diversity" in any of three ways. A small group of studies treat students' contacts with "diversity" largely as a function of the numerical or proportional racial/ethnic or gender mix of students on a campus (e.g., Chang, 1996, 1999a; Kanter, 1977; Sax, 1996) .... A second considerably larger set of studies take some modicum of structural diversity as a given and operationalizes students' encounters with diversity using the frequency or nature of their reported interactions with peers who are racially/ethnically different from themselves .... A third set of studies examines institutionally structured and purposeful programmatic efforts to help students engage racial/ethnic and/or gender "diversity" in the form of both ideas and people.

These various approaches have been used to examine the effects of diversity on a broad array of student educational outcomes. The evidence is almost uniformly consistent in indicating that students in a racial/ethnically or gender-diverse community, or engaged in a diversity-related activity, reap a wide array of positive educational benefits. *(Authors mention studies that have addressed the problem.)*

Only a relative handful of studies (e.g., Chang, 1996, 1999a; Sax, 1996) have specifically examined whether *the racial/ethnic or gender composition* of the students on a campus, in an academic major, or in a classroom (i.e., structural diversity) has the educational benefits claimed . . . . Whether the degree of racial diversity of a campus or classroom has a *direct* effect on learning outcomes, however, remains an open question. *(Deficiencies in the studies are noted.)*

The scarcity of information on the educational benefits of the structural diversity on a campus or in its classrooms is regrettable because it is the sort of evidence the courts appear to be requiring if they are to support race-sensitive admissions policies. *(Importance of the study for an audience mentioned.)*

This study attempted to contribute to the knowledge base by exploring the influence of structural diversity in the classroom on students' development of academic and intellectual skills. . . . This study examines both the direct effect of classroom diversity on academic/intellectual outcomes and whether any effects of classroom diversity may be moderated by the extent to which active and collaborative instructional approaches are used in the course. *(Purpose of the study is identified.)* (pp. 510–512, reprinted by permission of *The Journal of Higher Education*)

The Research Problem

In the Terenzini et al. (2001) article, the first sentence accomplishes both primary objectives for an introduction: piquing interest in the study and conveying a distinct research problem or issue. What effect did this sentence have? Would it entice a reader to read on? Was it pitched at a level so that a wide audience could understand it? These questions are important for opening sentences, and they are called a **narrative hook**, a term drawn from English composition, meaning words that serve to draw, engage, or hook the reader into the study. To learn how to write good narrative hooks, study opening sentences in leading journals in different fields of study. Often, journalists provide good examples in the lead sentences of newspaper and magazine articles. Here follow a few examples of lead sentences from social science journals.

- "The transsexual and ethnomethodological celebrity Agnes changed her identity nearly three years before undergoing sex reassignment surgery." (Cahill, 1989, p. 281)

- "Who controls the process of chief executive succession?" (Boeker, 1992, p. 400)

- "There is a large body of literature that studies the cartographic line (a recent summary article is Buttenfield 1985), and generalization of cartographic lines (McMaster 1987)." (Carstensen, 1989, p. 181)

All three of these examples present information easily understood by many readers. The first two—introductions in qualitative studies—demonstrate how reader interest can be created by reference to the single participant and by posing a question. The third example, a quantitative-experimental study, shows how one can begin with a literature perspective. All three examples demonstrate well how the lead sentence can be written so that the reader is not taken into a detailed morass of thought, but lowered gently into the topic.

I use the metaphor of the writer lowering a barrel into a well. The *beginning* writer plunges the barrel (the reader) into the depths of the well (the article). The reader sees only unfamiliar material. The *experienced* writer lowers the barrel (the reader, again) slowly, allowing the reader to acclimate to the depths (the study). This lowering of the barrel begins with *a narrative hook* of sufficient generality that the reader understands and can relate to the topic.

Beyond this first sentence, it is important to clearly identify the issue(s) or problem(s) that leads to a need for the study. Terenzini et al. (2001) discuss a distinct problem: the struggle to increase the racial and ethnic diversity on U.S. college and university campuses. They note that policies to increase diversity are at "the center of an intense national debate" (p. 509).

In applied social science research, problems arise from issues, difficulties, and current practices. The research problem in a study begins to become clear when the researcher asks, "What is the need for this study?" or "What problem influenced the need to undertake this study?" For example, schools may not have implemented multicultural guidelines, the needs of faculty in colleges are such that they need to engage in professional development activities in their departments, minority students need better access to universities, or a community needs to better understand the contributions of its early female pioneers. These are all significant research problems that merit further study and establish a practical issue or concern that needs to be addressed. When designing the opening paragraphs of a proposal, which includes the research problem, keep in mind these **research tips**:

● Write an opening sentence that will stimulate reader interest as well as convey an issue to which a broad audience can relate.

● As a general rule, refrain from using quotations, especially long ones, in the lead sentence. Quotations raise many possibilities for interpretation and thus create unclear beginnings. However, as is evident in some qualitative studies, quotations can create reader interest.

● Stay away from idiomatic expressions or trite phrases (e.g., "The lecture method remains a 'sacred cow' among most college and university instructors.").

● Consider numeric information for impact (e.g., "Every year, an estimated 5 million Americans experience the death of an immediate family member.").

● Clearly identify the research problem (i.e., dilemma, issue) leading to the study. Ask yourself, "Is there a specific sentence (or sentences) in which I convey the research problem?"

● Indicate why the problem is important by citing numerous references that justify the need to study the problem. In perhaps a not so joking manner, I say to my students that if they do not have a dozen references cited on the first page of their proposal, they do not have a scholarly study.

● Make sure that the problem is framed in a manner consistent with the approach to research in the study (e.g., exploratory in qualitative, examining relationships or predictors in quantitative, and either approach in mixed methods inquiry).

● Consider and write about whether there is a single problem involved in the proposed study or multiple problems that lead to a need for the study. Often, multiple research problems are addressed in research studies.

Studies Addressing the Problem

After establishing the research problem in the opening paragraphs, Terenzini et al. (2001) next justify its importance by reviewing studies that have examined the issue. I must be careful as I talk about reviewing studies here, because I do not have in mind a complete literature review for the introduction passage. It is later, in the literature review section of a proposal, that students thoroughly review the literature. Instead, in the introduction, this literature review passage should summarize large groups of studies instead of individual ones. I tell my students to reflect on their literature maps (described in Chapter 2) and look at and summarize the broad categories at the top into which they assigned their literature. Mentioning these broad categories are what I mean by reviewing studies in an introduction to a proposal.

The purpose of **reviewing studies** in an introduction is to justify the importance of the study and to create distinctions between past studies and the proposed one. This component might be called "setting the research problem within the ongoing dialogue in the literature." Researchers do not want to conduct a study that replicates exactly what someone else has studied. New studies need to add to the literature or to extend or retest what others have examined. Marshall and Rossman (2006) refer to this brief literature review in an introduction as a way to set the study within the context of other, related studies. The ability to frame the study in this way separates novices from more experienced researchers. The veteran has reviewed and understands what has been written about a topic or certain problem in the field. This knowledge comes from years of experience following the development of problems and their accompanying literature.

The question often arises as to what type of literature to review. My best advice would be to review research studies in which authors advance research questions and report data to answer them. These studies might be quantitative, qualitative, or mixed methods studies. The important point is that the literature provides studies about the research problem being addressed in the proposal. Beginning researchers often ask, "What do I do now? No research has been conducted on my topic." Of course, in some narrowly construed studies or in new, exploratory projects, no literature exists to document the research problem. Also, it makes sense that a topic is being proposed for study precisely because little research has been conducted on it. To counter this statement, I often suggest that an investigator think about the literature, using an inverted triangle as an image. At the apex of the inverted triangle lies the scholarly study being proposed. This study is narrow and focused (and studies may not exist on it). If one broadens the review of the literature upward to the base of the inverted triangle, literature can be found, although it may be somewhat removed from the study at hand. For example, the narrow topic of at-risk African Americans in primary school may not have been researched; however, more broadly speaking, the topic of at-risk students generally in the primary school or at any level in education, may have been studied. The researcher would summarize the more general literature and end with statements about a need for studies that examine at-risk African American students at the primary school level.

To review the literature related to the research problem for an introduction to a proposal, consider these **research tips**:

● Refer to the literature by summarizing groups of studies, not individual ones (unlike the focus on single studies in the integrated review in Chapter 2). The intent should be to establish broad areas of research.

● To deemphasize single studies, place the in-text references at the end of a paragraph or at the end of a summary point about several studies.

● Review research studies that used quantitative, qualitative, or mixed methods approaches.

● Find recent literature to summarize, such as that published in the past 10 years. Cite older studies if they are valuable because they have been widely referenced by others.

Deficiencies in Past Literature

After advancing the problem and reviewing the literature about it, the researcher then identifies *deficiencies* found in this literature. Hence, I call this template a *deficiencies model* for writing an introduction. The nature of these deficiencies varies from study to study. **Deficiencies in past literature** may exist because topics not have been explored with a particular group, sample, or population; the literature may need to be replicated or repeated to see if the same findings hold, given new samples of people or new sites for study; or the voice of underrepresented groups has not been heard in published literature. In any given study, authors may mention one or more of these deficiencies. Deficiencies can often be found in the "suggestions for future research" sections of journal articles, and authors can reference these ideas and provide further justification for their proposed study.

Beyond mentioning the deficiencies, proposal writers need to tell how their planned study will remedy or address these deficiencies. For example, because past studies have overlooked an important variable, a study will include it and analyze its effect: For instance, because past studies have overlooked the examination of Native Americans as a cultural group, a study will include them as the participants in the project.

In the two examples that follow, the authors point out the gaps or shortcomings of the literature. Notice their use of key phrases to indicate the shortcomings: "what remains to be explored," "little empirical research," and "very few studies."

Example 5.1 *Deficiencies in the Literature—Needed Studies*

For this reason, the meaning of war and peace has been explored extensively by social scientists (Cooper, 1965; Alvik, 1968; Rosell, 1968; Svancarova & Svancarova, 1967–68; Haavedsrud, 1970). What remains to be explored, however, is how veterans of past wars react to vivid scenes of a new war.

(Ziller, 1990, pp. 85–86)

> **Example 5.2**   *Deficiencies in the Literature—Few Studies*
>
> Despite an increased interest in micropolitics, it is surprising that so little empirical research has actually been conducted on the topic, especially from the perspectives of subordinates. Political research in educational settings is especially scarce: Very few studies have focused on how teachers use power to interact strategically with school principals and what this means descriptively and conceptually (Ball, 1987; Hoyle, 1986; Pratt, 1984).
>
> (Blase, 1989, p. 381)

In summary, when identifying deficiencies in the past literature, proposal developers might use the following **research tips**:

● Cite several deficiencies to make the case even stronger for a study.

● Identify specifically the deficiencies of other studies (e.g., methodological flaws, variables overlooked).

● Write about areas overlooked by past studies, including topics, special statistical treatments, significant implications, and so forth.

● Discuss how a proposed study will remedy these deficiencies and provide a unique contribution to the scholarly literature.

These deficiencies might be mentioned using a series of short paragraphs that identify three or four shortcomings of the past research or focus on one major shortcoming, as illustrated in the Terenzini et al. (2001) introduction.

Significance of a Study for Audiences

In dissertations, writers often include a specific section describing the **significance of the study** for select audiences, to convey the importance of the problem for different groups that may profit from reading and using the study. By including this section, the writer creates a clear rationale for the importance of the study. The more audiences that can be mentioned, the greater the importance of the study and the more it will be seen by readers to have wide application. In designing this section, one might include

● Three or four reasons that the study adds to the scholarly research and literature in the field

● Three or four reasons about how the study helps improve practice

● And three or four reasons as to why the study will improve policy.

In the example to follow, the author stated the significance of the study in the opening paragraphs of a journal article. This study by Mascarenhas (1989) examined ownership of industrial firms. He identified explicitly decision makers, organizational members, and researchers as the audience for the study.

Example 5.3   *Significance of the Study Stated in an Introduction to a Quantitative Study*

A study of an organization's ownership and its domain, defined here as markets served, product scope, customer orientation, and technology employed (Abell and Hammond, 1979; Abell, 1980; Perry and Rainey, 1988), is important for several reasons. First, understanding relationships among ownership and domain dimensions can help to reveal the underlying logic of organizations' activities and can help organization members evaluate strategies. . . . Second, a fundamental decision confronting all societies concerns the type of institutions to encourage or adopt for the conduct of activity. . . . Knowledge of the domain consequences of different ownership types can serve as input to that decision . . . . Third, researchers have often studied organizations reflecting one or two ownership types, but their findings may have been implicitly over generalized to all organizations.

(Mascarenhas, 1989, p. 582)

Terenzini et al. (2001) end their introduction by mentioning how courts could use the information of the study to require colleges and universities to support "race-sensitive admissions policies" (p. 512). In addition, the authors might have mentioned the importance of this study for admissions offices and students seeking admission as well as the committees that review applications for admission.

Finally, good introductions to research studies end with a statement of the purpose or intent of the study. Terenzini et al. (2001) ended their introduction by conveying that they planned to examine the influence of structural diversity on student skills in the classroom.

SUMMARY

This chapter provides advice about composing and writing an introduction to a scholarly study. The first element is to consider how the introduction incorporates the research problems associated with quantitative, qualitative, or mixed methods research. Then, a five-part introduction is suggested as a model or template to use. Called *the deficiencies model*, it is based on first identifying the research problem (and including a narrative hook). Then it includes briefly reviewing the literature that has addressed the problem, indicating one or more deficiencies in the past literature and suggesting how the study will remedy these deficiencies. This section is followed by specifying the audiences that will profit from research on the problem, and the introduction ends with a purpose statement that sets forth the intent of the study.

Writing Exercises

1. Draft several examples of narrative hooks for the introduction to a study and share these with colleagues to determine if the hooks draw readers in, create interest in the study, and are presented at a level to which readers can relate.

2. Write the introduction to a proposed study. Include one paragraph each for the research problem, the related literature about this problem, the deficiencies in the literature, and the audiences who will potentially find the study of interest.

3. Locate several research studies published in scholarly journals in a particular field of study. Review the introductions and locate the sentence or sentences in which the authors state the research problem or issue.

ADDITIONAL READINGS

Bem, D. J. (1987). Writing the empirical journal article. In M. P. Zanna & J. M. Darley (Eds.), *The compleat academic: A practical guide for the beginning social scientist* (pp. 171–201). New York: Random House.

Daryl Bem emphasizes the importance of the opening statement in published research. He provides a list of rules of thumb for opening statements, stressing the need for clear, readable prose and a structure that leads the reader step by step to the problem statement. Examples are provided of both satisfactory and unsatisfactory opening statements. Bem calls for opening statements that are accessible to the nonspecialist yet not boring to the technically sophisticated reader.

Maxwell, J. A. (2005). *Qualitative research design: An interactive approach* (2nd ed.). Thousand Oaks, CA: Sage.

Joe Maxwell reflects on the purpose of a proposal for a qualitative dissertation. One of the fundamental aspects of a proposal is to justify the project—to help readers understand not only what you plan to do but also why. He mentions the importance of identifying the issues you plan to address and indicating why they are important to study. In an example of a graduate student dissertation proposal, he shares the major issues the student has addressed to create an effective argument for the study.

Wilkinson, A. M. (1991). *The scientist's handbook for writing papers and dissertations.* Englewood Cliffs, NJ: Prentice Hall.

Antoinette Wilkinson identifies the three parts of an introduction: the derivation and statement of the problem and a discussion of its nature, the discussion of the background of the problem, and the statement of the research question. Her book offers numerous examples of these three parts, together with a discussion of how to write and structure an introduction. Emphasis is placed on ensuring that the introduction leads logically and inevitably to a statement of the research question.

CHAPTER SIX

The Purpose Statement

The last section of an introduction, as mentioned in Chapter 5, is to present a purpose statement that establishes the intent of the entire research study. It is the most important statement in the entire study, and it needs to be clearly and specifically presented. From it, all other aspects of the research follow. In journal articles, researchers write the purpose statement into introductions; in dissertations and dissertation proposals, it often stands as a separate section.

In this chapter devoted exclusively to the purpose statement, I address the reasons for developing it , key principles to use in its design, and examples of good models to use in crafting one for your proposal.

SIGNIFICANCE AND MEANING OF A PURPOSE STATEMENT

According to Locke et al. (2007), the purpose statement indicates "why you want to do the study and what you intend to accomplish" (p. 9). Unfortunately, proposal-writing texts give little attention to the purpose statement, and writers on method often incorporate it into discussions about other topics, such as specifying research questions or hypotheses. Wilkinson (1991), for example, refers to it within the context of the research question and objective. Other authors frame it as an aspect of the research problem (Castetter & Heisler, 1977). Closely examining their discussions, however, indicates that they both refer to the purpose statement as the central, controlling idea in a study.

This passage is called the *purpose statement* because it conveys the overall intent of a proposed study in a sentence or several sentences. In proposals, researchers need to distinguish clearly between the purpose statement, the research problem, and the research questions. The purpose statement sets forth the intent of the study, not the problem or issue leading to a need for the study (see Chapter 5). The purpose is also not the research questions, those questions that the data collection will attempt to answer (discussed in Chapter 7). Instead and again, the **purpose statement** sets the objectives, the intent, or the major idea of a proposal or a study. This idea builds on a need (the problem) and is refined into specific questions (the research questions).

Given the importance of the purpose statement, it is helpful to set it apart from other aspects of the proposal or study and to frame it as a single sentence or paragraph that readers can identify easily. Although qualitative, quantitative, and mixed methods purpose statements share similar topics, each is identified in the following paragraphs and illustrated with fill-in scripts for constructing a thorough but manageable purpose statement.

A Qualitative Purpose Statement

A good **qualitative purpose statement** contains information about the central phenomenon explored in the study, the participants in the study, and the research site. It also conveys an emerging design and uses research words drawn from the language of qualitative inquiry (Schwandt, 2007). Thus, one might consider several basic design features for writing this statement:

● Use words such as *purpose, intent,* or *objective* to signal attention to this statement as the central controlling idea. Set the statement off as a separate sentence or paragraph, and use the language of research, such as "The purpose (or intent or objective) of this study is (was) (will be) . . . . " Researchers often use the present or past verb tense in journal articles and dissertations, and the future tense in proposals, because researchers are presenting a plan for a study not yet undertaken.

● Focus on a single phenomenon (or concept or idea). Narrow the study to one idea to be explored or understood. This focus means that a purpose does not convey relating two or more variables or comparing two or more groups, as is typically found in quantitative research. Instead, advance a single phenomenon, recognizing that the study may develop into an exploration of relationships or comparisons among ideas. None of these related explorations can be anticipated at the beginning. For example, a project might begin by exploring chairperson roles in enhancing faculty development (Creswell & Brown, 1992). Other qualitative studies might start by exploring teacher identity and the marginalization of this identity for a teacher in a particular school (Huber & Whelan, 1999), the meaning of baseball culture in a study of the work and talk of stadium employees (Trujillo, 1992), or how individuals cognitively represent AIDS (Anderson & Spencer, 2002). These examples all illustrate the focus on a single idea.

● Use action verbs to convey how learning will take place. Action verbs and phrases, such as *describe, understand, develop, examine the meaning of,* or *discover,* keep the inquiry open and convey an emerging design.

● Use neutral words and phrases—nondirectional language—such as, exploring the "experiences of individuals" rather than the "successful experiences of individuals." Other words and phrases that may be problematic include *useful, positive*, and *informing*—all words that suggest an outcome that may or may not occur. McCracken (1988) refers to the need in qualitative interviews to let the respondent describe his or her experience. Interviewers (or purpose statement writers) can easily violate the "law of nondirection" (McCracken, 1988, p. 21) in qualitative research by using words that suggest a directional orientation.

● Provide a general working definition of the central phenomenon or idea, especially if the phenomenon is a term that is not typically understood by a broad audience. Consistent with the rhetoric of qualitative research, this definition is not rigid and set, but tentative and evolving throughout a study based on information from participants. Hence, a writer might use the words, "A tentative definition at this time for _____ (central phenomenon) is . . . . " It should also be noted that this definition is not to be confused with the detailed definition of terms section as discussed in Chapter 2 on the review of the literature. The intent here is to convey to readers at an early stage in a proposal or research study a general sense of the central phenomenon so that they can better understand information that unfolds during the study.

● Include words denoting the strategy of inquiry to be used in data collection, analysis, and the process of research, such as whether the study will use an ethnographic, grounded theory, case study, phenomenological, narrative approach, or some other strategy.

● Mention the participants in the study, such as whether there might be one or more individuals, a group of people, or an entire organization.

● Identify the site for the research, such as homes, classrooms, organizations, programs, or events. Describe this site in enough detail so that the reader knows exactly where a study will take place.

● As a final thought in the purpose statement, include some language that delimits the scope of participation or research sites in the study. For example, the study may be limited to women only, or Hispanics only. The research site may be limited to one metropolitan city or to one small geographic area. The central phenomenon may be limited to individuals in business organizations who participate in creative teams. These delimitations help to further define the parameters of the research study.

Although considerable variation exists in the inclusion of these points in purpose statements, a good dissertation or thesis proposal should contain many of them.

To assist you, here follows a script that should be helpful in drafting a complete statement. A **script**, as used in this book, contains the major words and ideas of a statement and provides space for the researcher to insert information.

> The purpose of this _____ (strategy of inquiry, such as ethnography, case study, or other type) study is (was? will be?) to _____ (understand? describe? develop? discover?) the _____ (central phenomenon being studied) for _____ (the participants, such as the individual, groups, organization) at _____ (research site). At this stage in the research, the _____ (central phenomenon being studied) will be generally defined as _____ (provide a general definition).

The following examples may not illustrate perfectly all the elements of this script, but they represent adequate models to study and emulate.

Example 6.1   *A Purpose Statement in a Qualitative Phenomenology Study*

Lauterbach (1993) studied five women who had each lost a baby in late pregnancy and their memories and experiences of this loss. Her purpose statement was as follows:

> The phenomenological inquiry, as part of uncovering meaning, articulated "essences" of meaning in mothers' lived experiences when their wished-for babies died. Using the lens of the feminist perspective, the focus was on mothers' memories and their "living through" experience. This perspective facilitated breaking through the silence surrounding mothers' experiences; it assisted in articulating and amplifying mothers' memories and their stories of loss. Methods of inquiry included phenomenological reflection on data elicited by existential investigation of mothers' experiences, and investigation of the phenomenon in the creative arts.
>
> (Lauterbach, 1993, p. 134)

I found Lauterbach's (1993) purpose statement in the opening section of the journal article under the heading, "Aim of Study." Thus, the heading calls attention to this statement. "Mothers' lived experiences" would be the central phenomenon, and the author uses the action word *portray* to discuss the *meaning* (a neutral word) of these experiences. The author further defined what experiences were examined when she identifies "memories" and "lived through" experiences. Throughout this passage, it is clear that Lauterbach used the strategy of phenomenology. Also, the passage conveys that the participants were mothers, and later in the article, the reader learns that the author interviewed a convenience sample of five mothers, each of whom had experienced a perinatal death of a child in her home.

Example 6.2   *A Purpose Statement in a Case Study*

Kos (1991) conducted a multiple case study of perceptions of reading-disabled middle-school students concerning factors that prevented these students from progressing in their reading development. Her purpose statement read as follows:

> The purpose of this study was to explore affective, social, and educational factors that may have contributed to the development of reading disabilities in four adolescents. The study also sought explanation as to why students' reading disabilities persisted despite years of instruction. This was not an intervention study and, although some students may have improved their reading, reading improvement was not the focus of the study.

> (Kos, 1991, pp. 876–877)

Notice Kos's (1991) disclaimer that this study was not a quantitative study measuring the magnitude of reading changes in the students. Instead, Kos clearly placed this study within the qualitative approach by using words such as "explore." She focused attention on the central phenomenon of "factors" and provided a tentative definition by mentioning examples, such as "affective, social, and educational." She included this statement under a heading called "Purpose of the Study" to call attention to it, and she mentioned the participants. In the abstract and the methodology section, a reader finds out that the study used the inquiry strategy of case study research and that the study took place in a classroom.

Example 6.3   *A Purpose Statement in an Ethnography*

Rhoads (1997) conducted a 2-year ethnographic study exploring how the campus climate can be improved for gay and bisexual males at a large university. His purpose statement, included in the opening section, was as follows:

> The article contributes to the literature addressing the needs of gay and bisexual students by identifying several areas where progress can be made in improving the campus climate for them. This paper derives from a two-year ethnographic study of a student subculture composed of gay and bisexual males at a large research university; the focus on men reflects the fact that lesbian and bisexual women constitute a separate student subculture at the university under study.

> (Rhoads, 1997, p. 276)

With intent to improve the campus, this qualitative study falls into the genre of advocacy research as mentioned in Chapter 1. Also, these sentences occur at the beginning of the article to signal the reader about the purpose of the study. The needs of these students become the central phenomenon under study, and the author seeks to identify areas that can improve the climate for gays and bisexual males. The author also mentioned that the strategy of inquiry will be ethnographic and that the study will involve males (participants) at a large university (site). At this point, the author does not provide additional information about the exact nature of these needs or a working definition to begin the article. However, he does refer to identity and proffers a tentative meaning for that term in the next section of the study.

Example 6.4   *A Purpose Statement in a Grounded Theory Study*

Richie et al. (1997) conducted a qualitative study to develop a theory of the career development of 18 prominent, highly achieving African American Black and White women in the United States working in different occupational fields. In the second paragraph of this study, they stated the purpose:

> The present article describes a qualitative study of the career development of 18 prominent, highly achieving African-American Black and White women in the United States across eight occupational fields. Our overall aim in the study was to explore critical influences on the career development of these women, particularly those related to their attainment of professional success.
>
> (Richie et al., 1997, p. 133)

In this statement, the central phenomenon is career development, and the reader learns that the phenomenon is defined as critical influences in the professional success of the women. In this study, *success*, a directional word, serves to define the sample of individuals to be studied more than to limit the inquiry about the central phenomenon. The authors plan to explore this phenomenon, and the reader learns that the participants are all women, in different occupational groups. Grounded theory as a strategy of inquiry is mentioned in the abstract and later in the procedure discussion.

A Quantitative Purpose Statement

Quantitative purpose statements differ considerably from the qualitative models in terms of the language and a focus on relating or comparing variables or constructs. Recall from Chapter 3 the types of major variables: independent, mediating, moderating, and dependent.

The design of a **quantitative purpose statement** includes the variables in the study and their relationship, the participants, and the research site. It also includes language associated with quantitative research and the deductive testing of relationships or theories. A quantitative purpose statement begins with identifying the proposed major variables in a study (independent, intervening, dependent), accompanied by a visual model to clearly identify this sequence, and locating and specifying how the variables will be measured or observed. Finally, the intent of using the variables quantitatively will be either to relate variables, as one typically finds in a survey, or to compare samples or groups in terms of an outcome, as commonly found in experiments.

The major components of a good quantitative purpose statement include the following:

● Include words to signal the major intent of the study, such as *purpose, intent,* or *objective.* Start with "The purpose (or objective or intent) of this study is (was, will be) . . ."

● Identify the theory, model, or conceptual framework. At this point one does not need to describe it in detail; in Chapter 3, I suggested the possibility of writing a separate "Theoretical Perspective" section for this purpose. Mentioning it in the purpose statement provides emphasis on the importance of the theory and foreshadows its use in the study.

● Identify the independent and dependent variables, as well as any mediating, moderating, or control variables used in the study.

● Use words that connect the independent and dependent variables to indicate that they are related, such as "the relationship between" two or more variables, or a "comparison of" two or more groups. Most quantitative studies employ one of these two options for connecting variables in the purpose statement. A combination of comparing and relating might also exist, for example, a two-factor experiment in which the researcher has two or more treatment groups as well as a continuous independent variable. Although one typically finds studies about comparing two or more groups in experiments, it is also possible to compare groups in a survey study.

● Position or order the variables from left to right in the purpose statement, with the independent variable followed by the dependent variable. Place intervening variables between the independent and dependent variables. Many researchers also place the moderating variables between the independent and dependent variables. Alternatively, control variables might be placed immediately following the dependent variable, in a phrase such as "controlling for . . ." In experiments, the independent variable will always be the manipulated variable.

● Mention the specific type of strategy of inquiry (such as survey or experimental research) used in the study. By incorporating this information, the researcher anticipates the methods discussion and enables a reader to associate the relationship of variables to the inquiry approach.

● Make reference to the participants (or the unit of analysis) in the study and mention the research site.

● Generally define each key variable, preferably using set and accepted established definitions found in the literature. General definitions are included at this point to help the reader best understand the purpose statement. They do not replace specific, operational definitions found later when a writer has a "Definition of Terms" section in a proposal (details about how variables will be measured). Also delimitations that affect the scope of the study might be mentioned, such as the scope of the data collection or limited to certain individuals.

Based on these points, a quantitative purpose statement script can include these ideas:

> The purpose of this _____ (experiment? survey?) study is (was? will be?) to test the theory of _____ that _____ (compares? relates?) the _____ (independent variable) to _____ (dependent variable), controlling for _____ (control variables) for _____ (participants) at _____ (the research site). The independent variable(s) _____ will be defined as _____ (provide a definition). The dependent variable(s) will be defined as _____ (provide a definition), and the control and intervening variable(s), _____, (identify the control and intervening variables) will be defined as _____ (provide a definition).

The examples to follow illustrate many of the elements in these scripts. The first two studies are surveys; the last one is an experiment.

Example 6.5 *A Purpose Statement in a Published Survey Study*

Kalof (2000) conducted a 2-year longitudinal study of 54 college women about their attitudes and experiences with sexual victimization. These women responded to two identical mail surveys administered 2 years apart. The author combined the purpose statement, introduced in the opening section, with the research questions.

> This study is an attempt to elaborate on and clarify the link between women's sex role attitudes and experiences with sexual victimization. I used

two years of data from 54 college women to answer these questions: (1) Do women's attitudes influence vulnerability to sexual coercion over a two-year period? (2) Are attitudes changed after experiences with sexual victimization? (3) Does prior victimization reduce or increase the risk of later victimization?

(Kalof, 2000, p. 48)

Although Kalof (2000) does not mention a theory that she seeks to test, she identifies both her independent variable (sex role attitudes) and the dependent variable (sexual victimization). She positioned these variables from independent to dependent. She also discussed linking rather than relating the variables to establish a connection between them. This passage identifies the participants (women) and the research site (a college setting). Later, in the method section, she mentioned that the study was a mailed survey. Although she does not define the major variables, she provides specific measures of the variables in the research questions.

Example 6.6   *A Purpose Statement in a Dissertation Survey Study*

DeGraw (1984) completed a doctoral dissertation in the field of education on the topic of educators working in adult correctional institutions. Under a section titled "Statement of the Problem," he advanced the purpose of the study:

The purpose of this study was to examine the relationship between personal characteristics and the job motivation of certified educators who taught in selected state adult correctional institutions in the United States. Personal characteristics were divided into background information about the respondent (i.e., institutional information, education level, prior training, etc.) and information about the respondents' thoughts of changing jobs. The examination of background information was important to this study because it was hoped it would be possible to identify characteristics and factors contributing to significant differences in mobility and motivation. The second part of the study asked the respondents to identify those motivational factors of concern to them. Job motivation was defined by six general factors identified in the educational work components study (EWCS) questionnaire (Miskel & Heller, 1973). These six factors are: potential for personal challenge and development; competitiveness; desirability and reward of success; tolerance for work pressures; conservative security; and willingness to seek reward in spite of uncertainty vs. avoidance.

(DeGraw, 1984, pp. 4, 5)

This statement included several components of a good purpose statement. It was presented in a separate section, it used the word *relationship*, terms were defined, and the population was mentioned. Further, from the order of the variables in the statement, one can clearly identify the independent variable and the dependent variable.

Example 6.7   *A Purpose Statement in an Experimental Study*

Booth-Kewley, Edwards, and Rosenfeld (1992) undertook a study comparing the social desirability of responding to a computer version of an attitude and personality questionnaire to the desirability of completing a pencil-and-paper version. They replicated a study completed on college students that used an inventory, called Balanced Inventory of Desirable Responding (BIDR), composed of two scales, impression management (IM) and self-deception (SD). In the final paragraph of the introduction, they advance the purpose of the study:

> We designed the present study to compare the responses of Navy recruits on the IM and SD scales, collected under three conditions—with paper-and-pencil, on a computer with backtracking allowed, and on a computer with no backtracking allowed. Approximately half of the recruits answered the questionnaire anonymously and the other half identified themselves.

> (Booth-Kewley et al, 1992, p. 563)

This statement also reflected many properties of a good purpose statement. The statement was separated from other ideas in the introduction as a separate paragraph, it mentioned that a comparison would be made, and it identified the participants in the experiment (i.e., the unit of analysis). In terms of the order of the variables, the authors advanced them with the dependent variable first, contrary to my suggestion (still, the groups are clearly identified). Although the theory base is not mentioned, the paragraphs preceding the purpose statement reviewed the findings of prior theory. The authors also do not tell us about the strategy of inquiry, but other passages, especially those related to procedures, discuss the study as an experiment.

A Mixed Methods Purpose Statement

A **mixed methods purpose statement** contains the overall intent of the study, information about both the quantitative and qualitative strands of the study, and a rationale of incorporating both strands to study the research problem. These statements need to be identified early, in the introduction, and they provide major signposts for the reader to understand the quantitative and qualitative parts of a study. Several guidelines might direct the organization and presentation of the mixed methods purpose statement:

● Begin with signaling words, such as "The purpose of " or "The intent of."

● Indicate the overall intent of the study from a content perspective, such as "The intent is to learn about organizational effectiveness" or "The intent is to examine families with step-children." In this way, the reader has an anchor to use to understand the overall study before the researcher divides the project into quantitative and qualitative strands.

● Indicate the type of mixed methods design, such as sequential, concurrent, or transformational, that will be used.

● Discuss the reasons for combining both quantitative and qualitative data. This reason could be one of the following (see Chapter 10 for more detail):

   • To better understand a research problem by converging (or triangulating) broad numeric trends from quantitative research and the detail of qualitative research

   • To explore participant views with the intent of building on these views with quantitative research so that they can be explored with a large sample of a population

   • To obtain statistical, quantitative results from a sample and then follow up with a few individuals to help explain those results in more depth (see also O'Cathain, Murphy, & Nicholl, 2007)

   • To best convey the trends and voices of marginalized groups or individuals

● Include the characteristics of a good qualitative purpose statement, such as focusing on a single phenomenon, using action words and nondirectional language, mentioning the strategy of inquiry, and identifying the participants and the research site.

● Include the characteristics of a good quantitative purpose statement, such as identifying a theory and the variables, relating variables or comparing groups in terms of variables, placing these variables in order from independent to dependent, mentioning the strategy of inquiry, and specifying the participants and research site for the research.

● Consider adding information about the specific types of both qualitative and quantitative data collection.

Based on these elements, four mixed methods purpose statement scripts follow (Creswell & Plano Clark, 2007). The first two are sequential studies with one type of data collection building on the other; the third is a concurrent study with both types of data collected at the same time and brought together in data analysis. The fourth example is a transformative mixed methods study script also based on a concurrent design.

1. A sequential study with a second quantitative phase building on an initial first qualitative phase:

   The intent of this two-phase, sequential mixed methods study is to _____ (mention content objective of the study). The first phase will be a qualitative exploration of a _____ (central phenomenon) by collecting _____ (types of data) from _____ (participants) at _____ (research site). Findings from this qualitative phase will then be used to test _____ (a theory, research questions, or hypotheses) that _____ (relate, compare) _____ (independent variable) with _____ (dependent variable) for _____ (sample of population) at _____ (research site). The reason for collecting qualitative data initially is that _____ (e.g., instruments are inadequate or not available, variables are not known, there is little guiding theory or few taxonomies).

2. A sequential study with the qualitative follow-up phase building on and helping to explain the initial quantitative phase:

   The intent of this two-phase, sequential mixed methods study will be to _____ (mention content objective of the study). In the first phase, quantitative research questions or hypotheses will address the _____ relationship or comparison of _____ (independent) and _____ (dependent) variables with _____ (participants) at _____ (the research site). Information from this first phase will be explored further in a second qualitative phase. In the second phase, qualitative interviews or observations will be used to probe significant _____ (quantitative results) by exploring aspects of the _____ (central phenomenon) with _____ (a few participants) at _____ (research site). The reason for following up with qualitative research in the second phase is to _____ (e.g., better understand and explain the quantitative results).

3. A concurrent study with the intent of gathering both quantitative and qualitative data and merging or integrating them to best understand a research problem:

   The intent of this concurrent mixed methods study is to _____ (content objective of the study). In the study, _____ (quantitative instruments) will be used to measure the relationship between _____ (independent variable) and _____ (dependent variable). At the same time, the _____ (central phenomenon) will be explored using _____ (qualitative interviews or observations) with

_____ (participants) at _____ (the research site). The reason for combining both quantitative and qualitative data is to better understand this research problem by converging both quantitative (broad numeric trends) and qualitative (detailed views) data.

4. This final script is one that a mixed methods researcher might use with a transformative mixed methods strategy of inquiry. The script is phrased for a concurrent study, but the mixed methods project might use either a concurrent (both quantitative and qualitative data collected at the same time) or a sequential (the two types of data collected in sequence or phases) strategy of inquiry. The elements that designate this script as transformational are that the intent of the study is to address an issue central to underrepresented or marginalized groups or individuals. Also, the outcome of such a study is to advocate for the needs of these groups or individuals, and this information is included in the purpose statement.

The intent of this concurrent mixed methods study is to _____ (state issue that needs to be addressed for group or individuals). In the study, _____ (quantitative instruments) will be used to measure the relationship between _____ (independent variable) and _____ (dependent variable). At the same time, the _____ (central phenomenon) will be explored using _____ (qualitative interviews or observations) with _____ (participants) at _____ (the research site). The reason for combining both quantitative and qualitative data is to better understand this research problem by converging both quantitative (broad numeric trends) and qualitative (detailed views) data and to advocate for change for _____ (groups or individuals).

Example 6.8   *A Concurrent Mixed Methods Purpose Statement*

Hossler and Vesper (1993) studied student and parent attitudes toward parental savings for the postsecondary education of their children. In this 3-year study, they identified the factors most strongly associated with parental savings and collected both quantitative and qualitative data. Their purpose statement was as follows:

In an effort to shed light on parental saving, this article examines parental saving behaviors. Using student and parent data from a longitudinal study employing multiple surveys over a three-year period, logistic regression was used to

(Continued)

(Continued)

identify the factors most strongly associated with parental savings for postsecondary education. In addition, insights gained from the interviews of a small subsample of students and parents who were interviewed five times during the three-year period are used to further examine parental savings.

(Hossler & Vesper, 1991, p. 141)

This section was contained under the heading "Purpose," and it indicated that both quantitative data (i.e., surveys) and qualitative data (i.e., interviews) were included in the study. Both forms of data were collected during the 3-year period, and the authors might have identified their study as a triangulation or concurrent design. Although the rationale for the study is not included in this passage, it is articulated later, in the methods discussion about surveys and interviews. Here we find that "the interviews were also used to explore variables under investigation in greater detail and triangulate findings using quantitative and qualitative data" (Hossler & Vesper, 1993, p. 146).

Example 6.9   *A Sequential Mixed Methods Purpose Statement*

Ansorge, Creswell, Swidler, and Gutmann (2001) studied the use of wireless iBook laptop computers in three teacher education methods courses. These laptop computers enabled students to work at their desks and use the laptops to log directly onto Web sites recommended by the instructors. The purpose statement was as follows:

The purpose of this sequential, mixed methods study was to first explore and generate themes about student use of iBook laptops in three teacher education classes using field observations and face-to-face interviews. Then, based on these themes, the second phase was to develop an instrument and to survey students about the laptop use on several dimensions. The rationale for using both qualitative and quantitative data was that a useful survey of student experience could best be developed only after a preliminary exploration of student use.

In this example, the statement begins with the signal words, "the purpose of." It then mentions the type of mixed methods design and contains the basic elements of both an initial qualitative phase and a follow-up quantitative phase. It includes information about both types of data collection and ends with a rationale for the incorporation of the two forms of data in a sequential design.

Example 6.10   *A Transformative Concurrent Mixed Methods Purpose Statement*

With this study we hope to contribute to the general understanding of how perceptions of fairness are formed and how gender equality is conceptualized by Swedish women and men. The aim of this article is to study the importance of time use, individual resources, distributive justice and gender ideology for perceptions of fairness and understandings of gender equality. Two Swedish studies are used to do this, a survey study and a qualitative interview study.

(Nordenmark & Nyman, 2003, p. 185)

This purpose statement begins with the intent of the study and presents the issue of gender equality as an issue of concern. This passage comes at the end of the introduction, and the reader has already learned that Sweden has a political goal of working toward gender equality in which "the balance of work and power between the sexes is eliminated" (Nordenmark & Nyman, 2003, p. 182). The authors mention the two types of data to be collected (survey and interviews), and following this passage, they refer to the advantages of combining both methods and that the two data sets will complement each other. Thus, a concurrent design is suggested. The purpose statement mentioned the quantitative variables that were related in the study; later we learn that several of these variables were also formed into qualitative research questions. However, the authors might have been more explicit in their quantitative and qualitative procedures as well as specifying the type of mixed methods strategy they used. Also, there was no mention in this purpose statement as to how this study would help create greater equality in Sweden. In the final section of the published study, however, the authors do suggest that conflicting goals and contradictory behavior and ideas may all impact gender equality in Sweden, and they call for measures of fairness and gender equality for large-scale surveys.

SUMMARY

This chapter emphasizes the primary importance of a purpose statement—it advances the central idea in a study. In writing a qualitative purpose statement, a researcher needs to identify a single central phenomenon and to pose a tentative definition for it. Also, the researcher employs action words, such as *discover, develop,* or *understand*; uses nondirectional language; and mentions the strategy of inquiry, the participants, and the research site. In a quantitative purpose statement, the researcher states the theory being tested as well as the variables and their relationship or comparison. It is important to position the independent variable first and the dependent variable second. The researcher conveys the strategy of inquiry as well as the participants and the research site for the investigation. In some purpose statements, the researcher also defines the key variables used in the study. In a mixed methods study, the type of strategy is mentioned as well as its rationale, such as whether the data are collected concurrently or sequentially. Many elements of both good qualitative and quantitative purpose statements are included.

Writing Exercises

1. Using the script for a qualitative purpose statement, write a statement by completing the blanks. Make this statement short; write no more than approximately three-quarters of a typed page.

2. Using the script for a quantitative purpose statement, write a statement. Also make this statement short, no longer than three-quarters of a typed page.

3. Using the script for a mixed methods purpose statement, write a purpose statement. Be sure to include the reason for mixing quantitative and qualitative data and incorporate the elements of both a good qualitative and a good quantitative purpose statement.

ADDITIONAL READINGS

Marshall, C., & Rossman, G. B. (2006). *Designing qualitative research* (4th ed.). Thousand Oaks, CA: Sage.

Catherine Marshall and Gretchen Rossman call attention to the major intent of the study, the purpose of the study. This section is generally embedded in the discussion of the topic, and it is mentioned in a sentence or two. It tells the reader what the results of the research are likely to accomplish. The authors characterize purposes as exploratory, explanatory, descriptive, and emanicipatory. They also mention that the purpose statement includes the unit of analysis (e.g., individuals, dyads, or groups).

Creswell, J. W., & Plano Clark, V. L. (2007). *Designing and conducting mixed methods research*. Thousand Oaks, CA: Sage.

John W. Creswell and Vicki L. Plano Clark have authored an overview and introduction to mixed methods research that covers the entire process of research from writing an introduction, collecting data, analyzing data, and interpreting and writing mixed methods studies. In their chapter on the introduction, they discuss qualitative, quantitative, and mixed methods purpose statements. They provide scripts and examples for four types of mixed methods studies as well as overall guidelines for writing these statements.

Wilkinson, A. M. (1991). *The scientist's handbook for writing papers and dissertations.* Englewood Cliffs, NJ: Prentice Hall.

Antoinette Wilkinson calls the purpose statement the "immediate objective" of the research study. She states that the purpose of the objective is to answer the research question. Further, the objective of the study needs to be presented in the introduction, although it may be implicitly stated as the subject of the research, the paper, or the method. If stated explicitly, the objective is found at the end of the argument in the introduction; it might also be found near the beginning or in the middle, depending on the structure of the introduction.

CHAPTER SEVEN

Research Questions and Hypotheses

Investigators place signposts to carry the reader through a plan for a study. The first signpost is the purpose statement, which establishes the central direction for the study. From the broad, general purpose statement, the researcher narrows the focus to specific questions to be answered or predictions based on hypotheses to be tested. This chapter begins by advancing several principles in designing and scripts for writing qualitative research questions; quantitative research questions, objectives, and hypotheses; and mixed methods research questions.

QUALITATIVE RESEARCH QUESTIONS

In a qualitative study, inquirers state research questions, not objectives (i.e., specific goals for the research) or hypotheses (i.e., predictions that involve variables and statistical tests). These research questions assume two forms: a central question and associated subquestions.

The **central question** is a broad question that asks for an exploration of the central phenomenon or concept in a study. The inquirer poses this question, consistent with the emerging methodology of qualitative research, as a general issue so as to not limit the inquiry. To arrive at this question, *ask*, "What is the broadest question that I can ask in the study?" Beginning researchers trained in *quantitative* research might struggle with this approach because they are accustomed to the reverse approach: identifying specific, narrow questions or hypotheses based on a few variables. In qualitative research, the intent is to explore the complex set of factors surrounding the central phenomenon and present the varied perspectives or meanings that participants hold. The following are guidelines for writing broad, qualitative research questions:

● *Ask one or two central questions followed by no more than five to seven subquestions*. Several subquestions follow each general central question; the subquestions narrow the focus of the study but leave open the questioning. This approach is well within the limits set by Miles and Huberman (1994), who recommended that researchers write no more than a dozen qualitative research questions in all (central and subquestions). The subquestions, in turn, can become specific questions used during interviews (or in observing or when looking at documents). In developing an interview protocol or guide, the researcher might ask an ice breaker question at the beginning, for example, followed by five or so subquestions in the study (see Chapter 9). The interview would then end with an additional wrap-up or summary question or ask, as I did in one of my qualitative case studies, "Who should I turn to, to learn more about this topic?" (Asmussen & Creswell, 1995).

● *Relate the central question to the specific qualitative strategy of inquiry.* For example, the specificity of the questions in ethnography at this stage of the design differs from that in other qualitative strategies. In ethnographic research, Spradley (1980) advanced a taxonomy of ethnographic questions that included a mini-tour of the culture-sharing group, their experiences, use of native language, contrasts with other cultural groups, and questions to verify the accuracy of the data. In critical ethnography, the research questions may build on a body of existing literature. These questions become working guidelines rather than truths to be proven (Thomas, 1993, p. 35). Alternatively, in phenomenology, the questions might be broadly stated without specific reference to the existing literature or a typology of questions. Moustakas (1994) talks about asking what the participants experienced and the contexts or situations in which they experienced it. A phenomenological example is, "What is it like for a mother to live with a teenage child who is dying of cancer?" (Nieswiadomy, 1993, p. 151). In grounded theory, the questions may be directed toward generating a theory of some process, such as the exploration of a process as to how caregivers and patients interact in a hospital setting. In a qualitative case study, the questions may address a description of the case and the themes that emerge from studying it.

● *Begin the research questions with the words* what *or* how *to convey an open and emerging design.* The word *why* often implies that the researcher is trying to explain why something occurs, and this suggests to me a cause-and-effect type of thinking that I associate with *quantitative* research instead of the more open and emerging stance of qualitative research.

● Focus on a single phenomenon or concept. As a study develops over time, factors will emerge that may influence this single phenomenon, but begin a study with a single focus to explore in great detail.

● Use exploratory verbs that convey the language of emerging design. These verbs tell the reader that the study will

- Discover (e.g., grounded theory)
- Seek to understand (e.g., ethnography)

- Explore a process (e.g., case study)
- Describe the experiences (e.g., phenomenology)
- Report the stories (e.g., narrative research)

● Use these more exploratory verbs that are nondirectional rather than directional words that suggest *quantitative* research, such as "affect," "influence," "impact," "determine," "cause," and "relate."

● Expect the research questions to evolve and change during the study in a manner consistent with the assumptions of an emerging design. Often in *qualitative* studies, the questions are under continual review and reformulation (as in a grounded theory study). This approach may be problematic for individuals accustomed to *quantitative* designs, in which the research questions remain fixed throughout the study.

● *Use open-ended questions* without reference to the literature or theory unless otherwise indicated by a qualitative strategy of inquiry.

● *Specify the participants and the research site* for the study, if the information has not yet been given.

Here is a script for a qualitative central question:

_____ (How or what) is the _____ ("story for" for narrative research; "meaning of" the phenomenon for phenomenology; "theory that explains the process of" for grounded theory; "culture-sharing pattern" for ethnography; "issue" in the "case" for case study) of _____ (central phenomenon) for _____ (participants) at _____ (research site).

The following are examples of qualitative research questions drawn from several types of strategies.

Example 7.1   *A Qualitative Central Question From an Ethnography*

Finders (1996) used ethnographic procedures to document the reading of teen magazines by middle-class European American seventh-grade girls. By examining the reading of teen zines (magazines), the researcher explored how the girls perceive and construct their social roles and relationships as they enter junior high school. She asked one guiding central question in her study:

How do early adolescent females read literature that falls outside the realm of fiction?

(Finders, 1996, p. 72)

Finders's (1996) central question begins with *how*; it uses an open-ended verb, *read*; it focuses on a single concept, the literature or teen magazines; and it mentions the participants, adolescent females, as the culture-sharing group. Notice how the author crafted a concise, single question that needed to be answered in the study. It is a broad question stated to permit participants to share diverse perspectives about reading the literature.

Example 7.2   *Qualitative Central Questions From a Case Study*

Padula and Miller (1999) conducted a multiple case study that described the experiences of women who went back to school, after a time away, in a psychology doctoral program at a major Midwestern research university. The intent was to document the women's experiences, providing a gendered and feminist perspective for women in the literature. The authors asked three central questions that guided the inquiry:

(a) How do women in a psychology doctoral program describe their decision to return to school? (b) How do women in a psychology doctoral program describe their reentry experiences? And (c) How does returning to graduate school change these women's lives?

(Padula & Miller, 1999, p. 328)

These three central questions all begin with the word *how*; they include open-ended verbs, such as "describe," and they focus on three aspects of the doctoral experience—returning to school, reentering, and changing. They also mention the participants as women in a doctoral program at a Midwestern research university.

QUANTITATIVE RESEARCH QUESTIONS AND HYPOTHESES

In quantitative studies, investigators use quantitative research questions and hypotheses, and sometimes objectives, to shape and specifically focus the purpose of the study. **Quantitative research questions** inquire about the relationships among variables that the investigator seeks to know. They are used frequently in social science research and especially in survey studies. **Quantitative hypotheses**, on the other hand, are predictions the researcher makes about the expected relationships among variables. They are numeric estimates of population values based on data collected from samples. Testing of hypotheses employs statistical procedures in which the investigator draws inferences about the population from a study sample. Hypotheses are used often in experiments in which investigators compare groups. Advisers often recommend their use in a formal research project, such as a dissertation or thesis, as a means of stating the direction a study will take. Objectives, on the other hand, indicate the goals or objectives for a study. They often appear in proposals for funding, but tend to be used with less frequency in social and health science research today. Because of this, the focus here will be on research questions and hypotheses. Here is an example of a script for a quantitative research question:

> Does _____ (name the theory) explain the relationship between _____ (independent variable) and _____ (dependent variable), controlling for the effects of _____ (control variable)?

Alternatively, a script for a quantitative null hypothesis might be as follows:

> There is no significant difference between _____ (the control and experimental groups on the independent variable) on _____ (dependent variable).

Guidelines for writing good quantitative research questions and hypotheses include the following.

● The use of variables in research questions or hypotheses is typically limited to three basic approaches. The researcher may *compare* groups on an independent variable to see its impact on a dependent variable. Alternatively, the investigator may *relate* one or more independent variables to one or more dependent variables. Third, the researcher may *describe* responses to the independent, mediating, or dependent variables. Most quantitative research falls into one or more of these three categories.

● The most rigorous form of quantitative research follows from a test of a theory (see Chapter 3) and the specification of research questions or hypotheses that are included in the theory.

● The independent and dependent variables must be measured separately. This procedure reinforces the cause-and-effect logic of quantitative research.

● To eliminate redundancy, write only research questions or hypotheses, not both, unless the hypotheses build on the research questions (discussion follows). Choose the form based on tradition, recommendations from an adviser or faculty committee, or whether past research indicates a prediction about outcomes.

● If hypotheses are used, there are two forms: null and alternative. A **null hypothesis** represents the traditional approach: it makes a prediction that in the general population, no relationship or no significant difference exists between groups on a variable. The wording is, "There is no difference (or relationship)" between the groups. The following example illustrates a null hypothesis.

Example 7.3   *A Null Hypothesis*

An investigator might examine three types of reinforcement for children with autism: verbal cues, a reward, and no reinforcement. The investigator collects behavioral measures assessing social interaction of the children with their siblings. A null hypothesis might read,

> There is no significant difference between the effects of verbal cues, rewards, and no reinforcement in terms of social interaction for children with autism and their siblings.

● The second form, popular in journal articles, is the alternative or **directional hypothesis**. The investigator makes a prediction about the expected outcome, basing this prediction on prior literature and studies on the topic that suggest a potential outcome. For example, the researcher may predict that "Scores will be higher for Group A than for Group B" on the dependent variable or that "Group A will change more than Group B" on the outcome. These examples illustrate a directional hypothesis because an expected prediction (e.g., higher, more change) is made. The following illustrates a directional hypothesis.

Example 7.4   *Directional Hypotheses*

Mascarenhas (1989) studied the differences between types of ownership (state-owned, publicly traded, and private) of firms in the offshore drilling industry. Specifically, the study explored such differences as domestic market dominance, international presence, and customer orientation. The study was a controlled field study using quasi-experimental procedures.

> Hypothesis 1: Publicly traded firms will have higher growth rates than privately held firms.

> Hypothesis 2: Publicly traded enterprises will have a larger international scope than state-owned and privately held firms.

Hypothesis 3: State-owned firms will have a greater share of the domestic market than publicly traded or privately held firms.

Hypothesis 4: Publicly traded firms will have broader product lines than state-owned and privately held firms.

Hypothesis 5: State-owned firms are more likely to have state-owned enterprises as customers overseas.

Hypothesis 6: State-owned firms will have a higher customer-base stability than privately held firms.

Hypothesis 7: In less visible contexts, publicly traded firms will employ more advanced technology than state-owned and privately held firms.

(Mascarenhas, 1989, pp. 585–588)

● Another type of alternative hypothesis is **nondirectional**—a prediction is made, but the exact form of differences (e.g., higher, lower, more, less) is not specified because the researcher does not know what can be predicted from past literature. Thus, the investigator might write, "There is a difference" between the two groups. An example follows which incorporates both types of hypotheses:

Example 7.5   Nondirectional and Directional Hypotheses

Sometimes directional hypotheses are created to examine the relationship among variables rather than to compare groups. For example, Moore (2000) studied the meaning of gender identity for religious and secular Jewish and Arab women in Israeli society. In a national probability sample of Jewish and Arab women, the author identified three hypotheses for study. The first is nondirectional and the last two are directional.

H_1: Gender identity of religious and secular Arab and Jewish women are related to different sociopolitical social orders that reflect the different value systems they embrace.

H_2: Religious women with salient gender identity are less socio-politically active than secular women with salient gender identities.

H_3: The relationships among gender identity, religiosity, and social actions are weaker among Arab women than among Jewish women.

● Unless the study intentionally employs demographic variables as predictors, use nondemographic variables (i.e., attitudes or behaviors) as independent and dependent variables. Because quantitative studies attempt to verify theories, demographic variables (e.g., age, income level, educational level, and so forth) typically enter these models as intervening (or mediating or moderating) variables instead of major independent variables.

● Use the same pattern of word order in the questions or hypotheses to enable a reader to easily identify the major variables. This calls for repeating key phrases and positioning the variables with the independent first and concluding with the dependent in left-to-right order (as discussed in Chapter 6 on good purpose statements). An example of word order with independent variables stated first in the phrase follows.

Example 7.6   *Standard Use of Language in Hypotheses*

1. There is no relationship between utilization of ancillary support services and academic persistence for non-traditional-aged women college students.

2. There is no relationship between family support systems and academic persistence for non-traditional-aged college women.

3. There is no relationship between ancillary support services and family support systems for non-traditional-aged college women.

A Model for Descriptive Questions and Hypotheses

Consider a model for writing questions or hypotheses based on writing descriptive questions (describing something) followed by inferential questions or hypotheses (drawing inferences from a sample to a population). These questions or hypotheses include both independent and dependent variables. In this model, the writer specifies descriptive questions for *each* independent and dependent variable and important intervening or moderating variables. Inferential questions (or hypotheses) that relate variables or compare groups follow these descriptive questions. A final set of questions may add inferential questions or hypotheses in which variables are controlled.

Example 7.7   *Descriptive and Inferential Questions*

To illustrate this approach, a researcher wants to examine the relationship of critical thinking skills (an independent variable measured on an instrument) to student achievement (a dependent variable measured by grades) in science classes for eighth-grade students in a large metropolitan school district. The researcher controls for the intervening effects of prior grades in science classes and parents' educational attainment. Following the proposed model, the research questions might be written as follows:

Descriptive Questions

1. How do the students rate on critical thinking skills? (A descriptive question focused on the independent variable)

2. What are the student's achievement levels (or grades) in science classes? (A descriptive question focused on the dependent variable)

3. What are the student's prior grades in science classes? (A descriptive question focused on the control variable of prior grades)

4. What is the educational attainment of the parents of the eighth-graders? (A descriptive question focused on another control variable, educational attainment of parents)

Inferential Questions

1. Does critical thinking ability relate to student achievement? (An inferential question relating the independent and the dependent variables)

2. Does critical thinking ability relate to student achievement, controlling for the effects of prior grades in science and the educational attainment of the eighth-graders' parents? (An inferential question relating the independent and the dependent variables, controlling for the effects of the two controlled variables)

This example illustrates how to organize all the research questions into descriptive and inferential questions. In another example, a researcher may want to compare groups, and the language may change to reflect this comparison in the inferential questions. In other studies, many more independent and dependent variables may be present in the model being tested, and a longer list of descriptive and inferential questions would result. I recommend this descriptive–inferential model.

This example also illustrates the use of variables to describe as well as relate. It specifies the independent variables in the first position in the questions, the dependent in the second, and the control variables in the third. It employs demographics as controls rather than central variables in the questions, and a reader needs to assume that the questions flow from a theoretical model.

MIXED METHODS RESEARCH QUESTIONS AND HYPOTHESES

In discussions about methods, researchers typically do not see specific questions or hypotheses especially tailored to mixed methods research. However, discussion has begun concerning the use of mixed methods questions in studies and also how to design them (see Creswell & Plano Clark, 2007; Tashakkori & Creswell, 2007). A strong mixed methods study should start with a mixed methods research question, to shape the methods and the overall design of a study. Because a mixed methods study relies on neither quantitative or qualitative research alone, some combination of the two provides the best information for the research questions and hypotheses. To be considered are what types of questions should be presented and when and what information is most needed to convey the nature of the study:

● Both qualitative and quantitative research questions (or hypotheses) need to be advanced in a mixed methods study in order to narrow and focus the purpose statement. These questions or hypotheses can be advanced at the beginning or when they emerge during a later phase of the research. For example, if the study begins with a quantitative phase, the investigator might introduce hypotheses. Later in the study, when the qualitative phase is addressed, the qualitative research questions appear.

● When writing these questions or hypotheses, follow the guidelines in this chapter for scripting good questions or hypotheses.

● Some attention should be given to the order of the research questions and hypotheses. In a two-phase project, the first-phase questions would come first, followed by the second-phase questions so that readers see them in the order in which they will be addressed in the proposed study. In a single-phase strategy of inquiry, the questions might be ordered according to the method that is given the most weight in the design.

● Include a **mixed methods research question** that directly addresses the mixing of the quantitative and qualitative strands of the research. This is the question that will be answered in the study based on the mixing (see Creswell & Plano Clark, 2007). This is a new form of question in research methods, and Tashakkori and Creswell (2007, p. 208) call it a "hybrid" or "integrated" question. This question could either be written at the beginning or when it emerges; for instance, in a two-phase study in which one phase builds on the other, the mixed methods questions might be placed in a discussion between the two phases. This can assume one of two forms. The first is to write it in a way that conveys the *methods* or *procedures* in a study (e.g., Does the qualitative data help explain the results from the initial quantitative phase of the study? See

Creswell & Plano Clark, 2007). The second form is to write it in a way that conveys the *content* of the study (e.g., Does the theme of social support help to explain why some students become bullies in schools? (see Tashakkori & Creswell, 2007.)

● Consider several different ways that all types of research questions (i.e., quantitative, qualitative, and mixed) can be written into a mixed methods study:

- Write separate quantitative questions or hypotheses and qualitative questions. These could be written at the beginning of a study or when they appear in the project if the study unfolds in stages or phases. With this approach, the emphasis is placed on the two approaches and not on the mixed methods or integrative component of the study.

- Write separate quantitative questions or hypotheses and qualitative questions and follow them with a mixed methods question. This highlights the importance of both the qualitative and quantitative phases of the study as well as their combined strength, and thus is probably the ideal approach.

- Write only a mixed methods question that reflects the *procedures* or the *content* (or write the mixed methods question in both a procedural and a content approach), and do not include separate quantitative and qualitative questions. This approach would enhance the viewpoint that the study intends to lead to some integration or connection between the quantitative and qualitative phases of the study (i.e., the sum of both parts is greater than each part).

Example 7.8   *Hypotheses and Research Questions in a Mixed Methods Study*

Houtz (1995) provides an example of a two-phase study with the separate quantitative and qualitative research hypotheses and questions stated in sections introducing each phase. She did not use a separate, distinct mixed methods research question. Her study investigated the differences between middle-school (nontraditional) and junior high (traditional) instructional strategies for seventh-grade and eighth-grade students and their attitudes toward science and their science achievement. Her study was conducted at a point when many schools were moving away from the two-year junior high concept to the three-year middle school (including sixth grade) approach to education. In this two-phase study, the first phase involved assessing pre-test

(Continued)

(Continued)

and post-test attitudes and achievement using scales and examination scores. Houtz then followed the quantitative results with qualitative interviews with science teachers, the school principal, and consultants. This second phase helped to explain differences and similarities in the two instructional approaches obtained in the first phase.

With a first-phase quantitative study, Houtz (1995, p. 630) mentioned the hypotheses guiding her research:

> It was hypothesized that there would be no significant difference between students in the middle school and those in the junior high in attitude toward science as a school subject. It was also hypothesized that there would be no significant difference between students in the middle school and those in the junior high in achievement in science.

These hypotheses appeared at the beginning of the study as an introduction to the quantitative phase. Prior to the qualitative phase, Houtz raised questions to explore the quantitative results in more depth. Focusing in on the achievement test results, she interviewed science teachers, the principal, and the university consultants and asked three questions:

> What differences currently exist between the middle school instructional strategy and the junior high instructional strategy at this school in transition? How has this transition period impacted science attitude and achievement of your students? How do teachers feel about this change process?
>
> (Houtz, 1995, p. 649)

Examining this mixed methods study shows that the author included both quantitative and qualitative questions, specified them at the beginning of each phase of her study, and used good elements for writing both quantitative hypotheses and qualitative research questions. Had Houtz (1995) developed a mixed methods question, it might have been stated from a *procedural* perspective:

> How do the interviews with teachers, the principal, and university consultants help to explain any quantitative differences in achievement for middle-school and junior high students?

Alternatively, the mixed methods question might have been written from a *content* orientation, such as:

> How do the themes mentioned by the teachers help to explain why middle-school children score lower than the junior high students?

Example 7.9   *A Mixed Methods Question Written in Terms of Mixing Procedures*

To what extent and in what ways do qualitative interviews with students and faculty members serve to contribute to a more comprehensive and nuanced understanding of this predicting relationship between CEEPT scores and student academic performance, via integrative mixed methods analysis?

(Lee & Greene, 2007)

This is a good example of a mixed methods question focused on the intent of mixing, to integrate the qualitative interviews and the quantitative data, the relationship of scores and student performance. This question emphasized what the integration was attempting to accomplish—a comprehensive and nuanced understanding—and at the end of the article, the authors presented evidence answering this question.

SUMMARY

Research questions and hypotheses narrow the purpose statement and become major signposts for readers. Qualitative researchers ask at least one central question and several subquestions. They begin the questions with words such as *how* or *what* and use exploratory verbs, such as *explore* or *describe*. They pose broad, general questions to allow the participants to explain their ideas. They also focus initially on one central phenomenon of interest. The questions may also mention the participants and the site for the research.

Quantitative researchers write either research questions or hypotheses. Both forms include variables that are described, related, categorized into groups for comparison, and the independent and dependent variables are measured separately. In many quantitative proposals, writers use research questions; however, a more formal statement of research employs hypotheses. These hypotheses are predictions about the outcomes of the results, and they may be written as alternative hypotheses specifying the exact results to be expected (more or less, higher or lower of something). They also may be stated in the null form, indicating no expected difference or no relationship between groups on a dependent variable. Typically, the researcher writes the independent variable(s) first, followed by the dependent variable(s). One model for ordering the questions in a quantitative proposal is to begin with descriptive questions followed by the inferential questions that relate variables or compare groups.

I encourage mixed methods researchers to construct separate mixed methods questions in their studies. This question might be written to emphasize the procedures or the content of the study, and it might be placed at different points. By writing this question, the researcher conveys the importance of integrating or combining the quantitative and qualitative elements. Several models exist for writing mixed methods questions into studies: writing only quantitative questions or hypotheses and qualitative questions, or writing both quantitative questions or hypotheses and qualitative questions followed by a mixed methods question, or writing only a mixed methods question.

Writing Exercises

1. For a qualitative study, write one or two central questions followed by five to seven subquestions.

2. For a quantitative study, write two sets of questions. The first set should be descriptive questions about the independent and dependent variables in the study. The second set should pose questions that relate (or compare) the independent variable(s) with the dependent variable(s). This follows the model presented in this chapter for combining descriptive and inferential questions.

3. Write a mixed methods research question. Write it first as a question incorporating the procedures of your mixed methods study and then rewrite it to incorporate the content. Comment on which approach works best for you.

ADDITIONAL READINGS

Creswell, J. W. (1999). Mixed-method research: Introduction and application. In G. J. Cizek (Ed.), *Handbook of educational policy* (pp. 455–472). San Diego: Academic Press.

In this chapter, I discuss the nine steps in conducting a mixed methods study. These are as follows:

1. Determine if a mixed methods study is needed to study the problem;

2. Consider whether a mixed methods study is feasible;

3. Write both qualitative and quantitative research questions;

4. Review and decide on the types of data collection;

5. Assess the relative weight and implementation strategy for each method;

6. Present a visual model;

7. Determine how the data will be analyzed;

8. Assess the criteria for evaluating the study; and

9. Develop a plan for the study.

In writing the research questions, I recommend developing both qualitative and quantitative types and stating within them the type of qualitative strategy of inquiry being used.

Tashakkori, A., & Creswell, J. W. (2007). Exploring the nature of research questions in mixed methods research. Editorial. *Journal of Mixed Methods Research, 1*(3), 207–211.

This editorial addresses the use and nature of research questions in mixed methods research. It highlights the importance of research questions in the process of research and discusses the need for a better understanding of the use of mixed methods questions. It asks, "How does one frame a research question in a mixed methods study?" (p. 207). Three models are presented: writing separate quantitative and qualitative questions, writing an overarching mixed methods question, or writing research questions for each phase of a study as the research evolves.

Morse, J. M. (1994). Designing funded qualitative research. In N. K. Denzin & Y. S. Lincoln (Eds.), *Handbook of qualitative research* (pp. 220–235). Thousand Oaks, CA: Sage.

Janice Morse, a nursing researcher, identifies and describes the major design issues involved in planning a qualitative project. She compares several strategies of inquiry and maps the type of research questions used in each strategy. For phenomenology and ethnography, the research calls for meaning and descriptive questions. For grounded theory, the questions need to address process, whereas in ethnomethodology and discourse analysis, the questions relate to verbal interaction and dialogue. She indicates that the wording of the research question determines the focus and scope of the study.

Tuckman, B. W. (1999). *Conducting educational research* (5th ed.). Fort Worth, TX: Harcourt Brace.

Bruce Tuckman provides an entire chapter on constructing hypotheses. He identifies the origin of hypotheses in deductive theoretical positions and in inductive observations. He further defines and illustrates both alternative and null hypotheses and takes the reader through the hypothesis testing procedure.

CHAPTER EIGHT

Quantitative Methods

For many proposal writers, the method section is the most concrete, specific part of a proposal. This chapter presents essential steps in designing quantitative methods for a research proposal or study, with specific focus on survey and experimental designs. These designs reflect postpositivist philosophical assumptions, as discussed in Chapter 1. For example, determinism suggests that examining the relationships between and among variables is central to answering questions and hypotheses through surveys and experiments. The reduction to a parsimonious set of variables, tightly controlled through design or statistical analysis, provides measures or observations for testing a theory. Objective data result from empirical observations and measures. Validity and reliability of scores on instruments lead to meaningful interpretations of data.

In relating these assumptions and the procedures that implement them, this discussion does not exhaustively treat quantitative research methods. Excellent, detailed texts provide information about survey research (e.g., see Babbie, 1990, 2007; Fink, 2002; Salant & Dillman, 1994). For experimental procedures, some traditional books (e.g., Campbell & Stanley, 1963; Cook & Campbell, 1979), as well as some newer texts, extend the ideas presented here (e.g., Bausell, 1994; Boruch, 1998; Field & Hole, 2003; Keppel, 1991; Lipsey, 1990; Reichardt & Mark, 1998). In this chapter, the focus is on the essential components of a method section in proposals for a survey and an experiment.

DEFINING SURVEYS AND EXPERIMENTS

A **survey design** provides a quantitative or numeric description of trends, attitudes, or opinions of a population by studying a sample of that population. From sample results, the researcher generalizes or makes claims about the population. In an *experiment*, investigators may also identify a sample and generalize to a population; however, the basic intent of an **experimental design** is to test the impact of a treatment (or an intervention) on an

outcome, controlling for all other factors that might influence that outcome. As one form of control, researchers randomly assign individuals to
groups. When one group receives a treatment and the other group does not,
the experimenter can isolate whether it is the treatment and not other factors that influence the outcome.

COMPONENTS OF A SURVEY METHOD PLAN

The design of a survey method section follows a standard format.
Numerous examples of this format appear in scholarly journals, and
these examples provide useful models. The following sections detail typical components. In preparing to design these components into a proposal,
consider the questions on the checklist shown in Table 8.1 as a general
guide.

The Survey Design

In a proposal or plan, one of the first parts of the method section can
introduce readers to the basic purpose and rationale for survey research.
Begin the discussion by reviewing the purpose of a survey and the rationale for its selection for the proposed study. This discussion can

● Identify the purpose of survey research. This purpose is to generalize
from a sample to a population so that inferences can be made about some
characteristic, attitude, or behavior of this population (Babbie, 1990).
Provide a reference to this purpose from one of the survey method texts
(several are identified in this chapter).

● Indicate why a survey is the preferred type of data collection procedure for the study. In this rationale, consider the advantages of survey
designs, such as the economy of the design and the rapid turnaround in
data collection. Discuss the advantage of identifying attributes of a large
population from a small group of individuals (Babbie, 1990; Fowler,
2002).

● Indicate whether the survey will be cross-sectional, with the data
collected at one point in time, or whether it will be longitudinal, with data
collected over time.

● Specify the form of data collection. Fink (2002) identifies four types:
self-administered questionnaires; interviews; structured record reviews
to collect financial, medical, or school information; and structured observations. The data collection may also involve creating a Web-based or
Internet survey and administering it online (Nesbary, 2000; Sue & Ritter,
2007). Regardless of the form of data collection, provide a rationale for the

Table 8.1	A Checklist of Questions for Designing a Survey Method
_____	Is the purpose of a survey design stated?
_____	Are the reasons for choosing the design mentioned?
_____	Is the nature of the survey (cross-sectional vs. longitudinal) identified?
_____	Are the population and its size mentioned?
_____	Will the population be stratified? If so, how?
_____	How many people will be in the sample? On what basis was this size chosen?
_____	What will be the procedure for sampling these individuals (e.g., random, nonrandom)?
_____	What instrument will be used in the survey? Who developed the instrument?
_____	What are the content areas addressed in the survey? The scales?
_____	What procedure will be used to pilot or field test the survey?
_____	What is the timeline for administering the survey?
_____	What are the variables in the study?
_____	How do these variables cross-reference with the research questions and items on the survey?
	What specific steps will be taken in data analysis to
(a)_____	analyze returns?
(b)_____	check for response bias?
(c)_____	conduct a descriptive analysis?
(d)_____	collapse items into scales?
(e)_____	check for reliability of scales?
(f)_____	run inferential statistics to answer the research questions?
_____	How will the results be interpreted?

procedure, using arguments based on its strengths and weaknesses, costs, data availability, and convenience.

The Population and Sample

Specify the characteristics of the population and the sampling procedure. Methodologists have written excellent discussions about the underlying logic of sampling theory (e.g., Babbie, 1990, 2007). Here are essential aspects of the population and sample to describe in a research plan:

● Identify the population in the study. Also state the size of this population, if size can be determined, and the means of identifying individuals in the population. Questions of access arise here, and the researcher might refer to availability of sampling frames—mail or published lists—of potential respondents in the population.

● Identify whether the sampling design for this population is single stage or multistage (called clustering). Cluster sampling is ideal when it is impossible or impractical to compile a list of the elements composing the population (Babbie, 2007). A single-stage sampling procedure is one in which the researcher has access to names in the population and can sample the people (or other elements) directly. In a multistage or clustering procedure, the researcher first identifies clusters (groups or organizations), obtains names of individuals within those clusters, and then samples within them.

● Identify the selection process for individuals. I recommend selecting a *random* sample, in which each individual in the population has an equal probability of being selected (a systematic or probabilistic sample). Less desirable is a nonprobability sample (or convenience sample), in which respondents are chosen based on their convenience and availability (Babbie, 1990). With randomization, a representative sample from a population provides the ability to generalize to a population.

● Identify whether the study will involve stratification of the population before selecting the sample. *Stratification* means that specific characteristics of individuals (e.g., both females and males) are represented in the sample and the sample reflects the true proportion in the population of individuals with certain characteristics (Fowler, 2002). When randomly selecting people from a population, these characteristics may or may not be present in the sample in the same proportions as in the population; stratification ensures their representation. Also identify the characteristics used in stratifying the population (e.g., gender, income levels, education). Within each stratum, identify whether the sample contains individuals with the characteristic in the same proportion as the characteristic appears in the entire population (Babbie, 1990; Miller, 1991).

● Discuss the procedures for selecting the sample from available lists. The most rigorous method for selecting the sample is to choose individuals using a random numbers table, a table available in many introductory statistics texts (e.g., Gravetter & Wallnau, 2000).

● Indicate the number of people in the sample and the procedures used to compute this number. In survey research, I recommend that one use a sample size formula available in many survey texts (e.g., see Babbie, 1990; Fowler, 2002).

Instrumentation

As part of rigorous data collection, the proposal developer also provides detailed information about the actual survey instrument to be used in the proposed study. Consider the following:

● Name the survey instrument used to collect data. Discuss whether it is an instrument designed for this research, a modified instrument, or an intact instrument developed by someone else. If it is a modified instrument, indicate whether the developer has provided appropriate permission to use it. In some survey projects, the researcher assembles an instrument from components of several instruments. Again, permission to use any part of other instruments needs to be obtained. In addition, instruments are being increasingly designed for online surveys (see Sue & Ritter, 2007). An online survey tool is SurveyMonkey (SurveyMonkey.com), a commercial product available since 1999. Using this service, researchers can create their own surveys quickly using custom templates and post them on Web sites or e-mail them for participants to complete. SurveyMonkey then can generate results and report them back to the researcher as descriptive statistics or as graphed information. The results can be downloaded into a spreadsheet or a database for further analysis. The basic program is free for 100 responses per survey and no more than 10 questions per survey. For additional responses, more questions, and several custom features, SurveyMonkey charges a monthly or annual fee.

● To use an existing instrument, describe the established validity and reliability of scores obtained from past use of the instrument. This means reporting efforts by authors to establish **validity**—whether one can draw meaningful and useful inferences from scores on the instruments. The three traditional forms of validity to look for are content validity (do the items measure the content they were intended to measure?), predictive or concurrent validity (do scores predict a criterion measure? Do results correlate with other results?), and construct validity (do items measure hypothetical constructs or concepts?). In more recent studies, construct validity has also included whether the scores serve a useful purpose and have positive consequences when they are used in practice (Humbley & Zumbo, 1996). Establishing the validity of the scores in a survey helps to identify whether an instrument might be a good one to use in survey research. This form of validity is different than identifying the threats to validity in experimental research, as discussed later in this chapter.

Also discuss whether scores resulting from past use of the instrument demonstrate **reliability**. Look for whether authors report measures of internal consistency (are the items' responses consistent across constructs?)

and test–retest correlations (are scores stable over time when the instrument is administered a second time?). Also determine whether there was consistency in test administration and scoring (were errors caused by carelessness in administration or scoring?; Borg, Gall, & Gall, 1993).

● When one modifies an instrument or combines instruments in a study, the original validity and reliability may not hold for the new instrument, and it becomes important to reestablish validity and reliability during data analysis.

● Include sample items from the instrument so that readers can see the actual items used. In an appendix to the proposal, attach sample items or the entire instrument.

● Indicate the major content sections in the instrument, such as the cover letter (Dillman, 1978, provides a useful list of items to include in cover letters), the items (e.g., demographics, attitudinal items, behavioral items, factual items), and the closing instructions. Also mention the type of scales used to measure the items on the instrument, such as continuous scales (e.g., *strongly agree* to *strongly disagree*) and categorical scales (e.g., yes/no, rank from highest to lowest importance).

● Discuss plans for pilot testing or field testing the survey and provide a rationale for these plans. This testing is important to establish the content validity of an instrument and to improve questions, format, and scales. Indicate the number of people who will test the instrument and the plans to incorporate their comments into final instrument revisions.

● For a mailed survey, identify steps for administering the survey and for following up to ensure a high response rate. Salant and Dillman (1994) suggest a four-phase administration process. The first mail-out is a short advance-notice letter to all members of the sample, and the second mail-out is the actual mail survey, distributed about 1 week after the advance-notice letter. The third mail-out consists of a postcard follow-up sent to all members of the sample 4 to 8 days after the initial questionnaire. The fourth mail-out, sent to all nonrespondents, consists of a personalized cover letter with a handwritten signature, the questionnaire, and a preaddressed return envelope with postage. Researchers send this fourth mail-out 3 weeks after the second mail-out. Thus, in total, the researcher concludes the administration period 4 weeks after its start, providing the returns meet project objectives.

Variables in the Study

Although readers of a proposal learn about the variables in purpose statements and research questions/hypotheses sections, it is useful in the method section to relate the variables to the specific questions or hypotheses on the instrument. One technique is to relate the variables, the research questions or hypotheses, and items on the survey instrument so that a reader can easily determine how the researcher will use the questionnaire items. Plan to include a table and a discussion that cross-reference the variables, the questions or hypotheses, and specific survey items. This procedure is especially helpful in dissertations in which investigators test large-scale models. Table 8.2 illustrates such a table using hypothetical data.

Table 8.2 Variables, Research Questions, and Items on a Survey		
Variable Name	**Research Question**	**Item on Survey**
Independent Variable 1: Prior publications	Descriptive research Question 1: How many publications did the faculty member produce prior to receipt of the doctorate?	See Questions 11, 12, 13, 14, and 15: publication counts for journal articles, books, conference papers, book chapters published before receiving the doctorate
Dependent Variable 1: Grants funded	Descriptive research Question 3: How many grants has the faculty member received in the past 3 years?	See Questions 16, 17, and 18: grants from foundations, federal grants, state grants
Control Variable 1: Tenure status	Descriptive research Question 5: Is the faculty member tenured?	See Question 19: tenured (yes/no)

Data Analysis and Interpretation

In the proposal, present information about the steps involved in analyzing the data. I recommend the following **research tips**, presenting them as a series of steps so that a reader can see how one step leads to another for a complete discussion of the data analysis procedures.

Step 1. Report information about the number of members of the sample who did and did not return the survey. A table with numbers and percentages describing respondents and nonrespondents is a useful tool to present this information.

Step 2. Discuss the method by which response bias will be determined. **Response bias** is the effect of nonresponses on survey estimates (Fowler, 2002). *Bias* means that if nonrespondents had responded, their responses would have substantially changed the overall results. Mention the procedures used to check for response bias, such as wave analysis or a respondent/nonrespondent analysis. In wave analysis, the researcher examines returns on select items week by week to determine if average responses change (Leslie, 1972). Based on the assumption that those who return surveys in the final weeks of the response period are nearly all nonrespondents, if the responses begin to change, a potential exists for response bias. An alternative check for response bias is to contact a few nonrespondents by phone and determine if their responses differ substantially from respondents. This constitutes a respondent-nonrespondent check for response bias.

Step 3. Discuss a plan to provide a **descriptive analysis** of data for all independent and dependent variables in the study. This analysis should indicate the means, standard deviations, and range of scores for these variables.

Step 4. If the proposal contains an instrument with scales or a plan to develop scales (combining items into scales), identify the statistical procedure (i.e., factor analysis) for accomplishing this. Also mention reliability checks for the internal consistency of the scales (i.e., the Cronbach alpha statistic).

Step 5. Identify the statistics and the statistical computer program for testing the major inferential research questions or hypotheses in the proposed study. The **inferential questions** or **hypotheses** relate variables or compare groups in terms of variables so that inferences can be drawn from the sample to a population. Provide a rationale for the choice of statistical test and mention the assumptions associated with the statistic. As shown in Table 8.3, base this choice on the nature of the research question (e.g., relating variables or comparing groups as the most popular), the number of independent and dependent variables, and the number of variables controlled (e.g., see Rudestam & Newton, 2007). Further, consider whether the variables will be measured on an instrument as a continuous score (e.g., age, from 18 to 36) or as a categorical score (e.g., women = 1, men = 2). Finally, consider whether the scores from the sample might be normally distributed in a bell-shaped curve if plotted out on a graph or non-normally distributed. There are additional ways to determine if the scores are normally distributed (see Creswell, 2008). These factors, in combination, enable a researcher to determine what statistical test will be suited for answering the research question or hypothesis. In Table 8.3, I show how the factors, in combination, lead to the selection of a number of common statistical tests. For further types of statistical tests, readers are referred to statistics methods books, such as Gravetter and Wallnau (2000).

Step 6. A final step in the data analysis is to present the results in tables or figures and interpret the results from the statistical test. An **interpretation of the results** means that the researcher draws conclusions from the results for the research questions, hypotheses, and the larger meaning of the results. This interpretation involves several steps.

Table 8.3 Criteria for Choosing Select Statistical Tests

Nature of Question	Number of Independent Variables	Number of Dependent Variables	Number of Control Variables (covariates)	Type of Score Independent/ Dependent Variables	Distribution of Scores	Statistical Test
Group comparison	1	1	0	Categorical/ continuous	Normal	t-test
Group comparison	1 or more	1	0	Categorical/ continuous	Normal	Analysis of variance
Group comparison	1 or more	1	1	Categorical/ continuous	Normal	Analysis of covariance
Group comparison	1	1	0	Categorical/ continuous	Non-normal	Mann-Whitney U test
Association between groups	1	1	0	Categorical/ categorical	Non-normal	Chi-square
Relate variables	1	1	0	Continuous/ continuous	Normal	Pearson product moment corrolation
Relate variables	2 or more	1	0	Continuous/ continuous	Normal	Multiple regression
Relate variables	1	1 or more	0	Categorical/ categorical	Non-normal	Spearman rank-order correlation

- Report whether the results of the statistical test were statistically significant or not, such as "the analysis of variance revealed a statistically significant difference between men and women in terms of attitudes toward banning smoking in restaurants $F(2; 6) = 8.55, p = .001$."

- Report how these results answered the research question or hypothesis. Did the results support the hypothesis or did they contradict what was expected?

- Indicate what might explain why the results occurred. This explanation might refer back to the theory advanced in the proposed study (see Chapter 3), past literature as reviewed in the literature review (see Chapter 2), or logical reasoning.

- Discuss the implications of the results for practice or for future research on the topic.

Example 8.1   *A Survey Method Section*

An example follows of a survey method section that illustrates many of the steps just mentioned. This excerpt (used with permission) comes from a journal article reporting a study of factors affecting student attrition in one small liberal arts college (Bean & Creswell, 1980, pp. 321–322).

Methodology

The site of this study was a small (enrollment 1,000), religious, coeducational, liberal arts college in a Midwestern city with a population of 175,000 people. *(Authors identified the research site and population.)*

The dropout rate the previous year was 25%. Dropout rates tend to be highest among freshmen and sophomores, so an attempt was made to reach as many freshmen and sophomores as possible by distribution of the questionnaire through classes. Research on attrition indicates that males and females drop out of college for different reasons (Bean, 1978, in press; Spady, 1971). Therefore, only women were analyzed in this study.

During April 1979, 169 women returned questionnaires. A homogeneous sample of 135 women who were 25 years old or younger, unmarried, full-time U.S. citizens, and Caucasian was selected for this analysis to exclude some possible confounding variables (Kerlinger, 1973).

Of these women, 71 were freshmen, 55 were sophomores, and 9 were juniors. Of the students, 95% were between the ages of 18 and 21. This sample is biased toward higher-ability students as indicated by scores on the ACT test. *(Authors presented descriptive information about the sample.)*

Data were collected by means of a questionnaire containing 116 items. The majority of these were Likert-like items based on a scale from "a very small extent" to "a very great extent." Other questions asked for factual information, such as ACT scores, high school grades, and parents' educational level. All information used in this analysis was derived from questionnaire data. This questionnaire had been developed and tested at three other institutions before its use at this college. *(Authors discussed the instrument.)*

Concurrent and convergent validity (Campbell & Fiske, 1959) of these measures was established through factor analysis, and was found to be at an adequate level. Reliability of the factors was established through the coefficient alpha. The constructs were represented by 25 measures—multiple items combined on the basis of factor analysis to make indices—and 27 measures were single item indicators. *(Validity and reliability were addressed.)*

Multiple regression and path analysis (Heise, 1969; Kerlinger & Pedhazur, 1973) were used to analyze the data.

In the causal model . . . , intent to leave was regressed on all variables which preceded it in the causal sequence. Intervening variables significantly related to intent to leave were then regressed on organizational variables, personal variables, environmental variables, and background variables. *(Data analysis steps were presented.)*

COMPONENTS OF AN EXPERIMENTAL METHOD PLAN

An experimental method discussion follows a standard form: partici-pants, materials, procedures, and measures. These four topics generally are sufficient. In this section of the chapter, I review these components as well as information about the experimental design and statistical analysis. As with the section on surveys, the intent here is to highlight key topics to be addressed in an experimental method proposal. An over-all guide to these topics is found by answering the questions on the checklist shown in Table 8.4.

Participants

Readers need to know about the selection, assignment, and number of participants who will take part in the experiment. Consider the following suggestions when writing the method section for an experiment:

● Describe the selection process for participants as either random or nonrandom (e.g., conveniently selected). The participants might be selected by *random selection* or *random sampling*. With random selection or **random sampling**, each individual has an equal probability of being selected from the population, ensuring that the sample will be representa-tive of the population (Keppel, 1991). In many experiments, however, only a *convenience* sample is possible because the investigator must use naturally formed groups (e.g., a classroom, an organization, a family unit) or volun-teers. When individuals are not randomly assigned, the procedure is called a **quasi-experiment.**

● When individuals can be randomly assigned to groups, the proce-dure is called a **true experiment.** If random assignment is made, discuss how the project will *randomly assign* individuals to the treatment groups. This means that of the pool of participants, Individual 1 goes to Group 1, Individual 2 to Group 2, and so forth, so that there is no systematic bias in assigning the individuals. This procedure eliminates the possibility of sys-tematic differences among characteristics of the participants that could affect the outcomes, so that any differences in outcomes can be attributed to the experimental treatment (Keppel, 1991).

● Identify other features in the experimental design that will system-atically control the variables that might influence the outcome. One approach is **matching participants** in terms of a certain trait or charac-teristic and then assigning one individual from each matched set to each group. For example, scores on a pre-test might be obtained. Individuals might then be assigned to groups, with each group having the same num-bers of high, medium, and low scorers on the pre-test. Alternatively, the criteria for matching might be ability levels or demographic variables.

Table 8.4	A Checklist of Questions for Designing an Experimental Procedure
_____	Who are the participants in the study?
_____	What is the population to which the results of the participants will be generalized?
_____	How were the participants selected? Was a random selection method used?
_____	How will the participants be randomly assigned? Will they be matched? How?
_____	How many participants will be in the experimental and control group(s)?
_____	What is the dependent variable or variables (i.e., outcome variable) in the study? How will it be measured? Will it be measured before and after the experiment?
_____	What is the treatment condition(s)? How was it operationalized?
_____	Will variables be covaried in the experiment? How will they be measured?
_____	What experimental research design will be used? What would a visual model of this design look like?
_____	What instrument(s) will be used to measure the outcome in the study? Why was it chosen? Who developed it? Does it have established validity and reliability? Has permission been sought to use it?
_____	What are the steps in the procedure (e.g., random assignment of participants to groups, collection of demographic information, administration of pretest, administration of treatment(s), administration of posttest)?
_____	What are potential threats to internal and external validity for the experimental design and procedure? How will they be addressed?
_____	Will a pilot test of the experiment be conducted?
_____	What statistics will be used to analyze the data (e.g., descriptive and inferential)?
_____	How will the results be interpreted?

A researcher may decide not to match, however, because it is expensive, takes time (Salkind, 1990), and leads to incomparable groups if participants leave the experiment (Rosenthal & Rosnow, 1991). Other procedures to place control into experiments involve using covariates (e.g., pre-test scores) as moderating variables and controlling for their effects statistically, selecting homogeneous samples, or blocking the participants into subgroups or categories and analyzing the impact of each subgroup on the outcome (Creswell, 2008).

● Tell the reader about the number of participants in each group and the systematic procedures for determining the size of each group. For experimental research, investigators use a power analysis (Lipsey, 1990) to identify the appropriate sample size for groups. This calculation involves

  • A consideration of the level of statistical significance for the experiment, or alpha

  • The amount of power desired in a study—typically presented as high, medium, or low—for the statistical test of the null hypothesis with sample data when the null hypothesis is, in fact, false

  • The effect size, the expected differences in the means between the control and experimental groups expressed in standard deviation units

● Researchers set values for these three factors (e.g., alpha = .05, power = .80, and effect size = .50) and can look up in a table the size needed for each group (see Cohen, 1977; Lipsey, 1990). In this way, the experiment is planned so that the size of each treatment group provides the greatest sensitivity that the effect on the outcome actually is due to the experimental manipulation in the study.

Variables

The variables need to be specified in an experiment so that it is clear to readers what groups are receiving the experimental treatment and what outcomes are being measured. Here are some suggestions for developing ideas about variables in a proposal:

● Clearly identify the *independent variables* in the experiment (recall the discussion of variables in Chapter 3). One independent variable must be the *treatment variable.* One or more groups receive the experimental manipulation, or treatment, from the researcher. Other independent variables may simply be measured variables in which no manipulation occurs (e.g., attitudes or personal characteristics of participants). Still other independent variables can be statistically controlled, such as demographics (e.g., gender or age). The method section must list and clearly identify all the independent variables in an experiment.

● Identify the *dependent variable or variables* (i.e., the outcomes) in the experiment. The dependent variable is the response or the criterion variable that is presumed to be caused by or influenced by the independent treatment conditions and any other independent variables). Rosenthal and

Rosnow (1991) advanced three prototypic outcomes measures: the direction of observed change, the amount of this change, and the ease with which the participant changes (e.g., the participant reacquires the correct response as in a single-subject design).

Instrumentation and Materials

During an experiment, one makes observations or obtains measures using instruments at a pre-test or post-test (or both) stage of the procedures. A sound research plan calls for a thorough discussion about the instrument or instruments—their development, their items, their scales, and reports of reliability and validity of scores on past uses. The researcher also should report on the materials used for the experimental treatment (e.g., the special program or specific activities given to the experimental group).

● Describe the instrument or instruments participants complete in the experiment, typically completed before the experiment begins and at its end. Indicate the established validity and reliability of the scores on instruments, the individuals who developed them, and any permissions needed to use them.

● Thoroughly discuss the materials used for the experimental treatment. One group, for example, may participate in a special computer-assisted learning plan used by a teacher in a classroom. This plan might involve handouts, lessons, and special written instructions to help students in this experimental group learn how to study a subject using computers. A pilot test of these materials may also be discussed, as well as any training required to administer the materials in a standard way. The intent of this pilot test is to ensure that materials can be administered without variability to the experimental group.

Experimental Procedures

The specific experimental design procedures also need to be identified. This discussion involves indicating the overall experiment type, citing reasons for the design, and advancing a visual model to help the reader understand the procedures.

● Identify the type of experimental design to be used in the proposed study. The types available in experiments are pre-experimental designs, true experiments, quasi-experiments, and single-subject designs. With *pre-experimental* designs, the researcher studies a single group and provides an intervention during the experiment. This design does not have a control group to compare with the experimental group. In *quasi-experiments,* the investigator uses control and experimental groups but does not randomly assign participants to groups (e.g., they may be intact groups available to the researcher). In a *true experiment,* the investigator randomly assigns the participants to treatment groups. A **single-subject design** or N of 1 design involves observing the behavior of a single individual (or a small number of individuals) over time.

● Identify what is being compared in the experiment. In many experiments, those of a type called *between-subject* designs, the investigator compares two or more groups (Keppel, 1991; Rosenthal & Rosnow, 1991). For example, a *factorial design* experiment, a variation on the between-group design, involves using two or more treatment variables to examine the independent and simultaneous effects of these treatment variables on an outcome (Vogt, 1999). This widely used behavioral research design explores the effects of each treatment separately and also the effects of variables used in combination, thereby providing a rich and revealing multidimensional view (Keppel, 1991). In other experiments, the researcher studies only one group in what is called a *within-group* design. For example, in a *repeated measures* design, participants are assigned to different treatments at different times during the experiment. Another example of a within-group design would be a study of the behavior of a single individual over time in which the experimenter provides and withholds a treatment at different times in the experiment, to determine its impact.

● Provide a diagram or a figure to illustrate the specific research design to be used. A standard notation system needs to be used in this figure. A **research tip** I recommend is to use a classic notation system provided by Campbell and Stanley (1963, p. 6):

- X represents an exposure of a group to an experimental variable or event, the effects of which are to be measured.

- O represents an observation or measurement recorded on an instrument.

- X's and O's in a given row are applied to the same specific persons. X's and O's in the same column, or placed vertically relative to each other, are simultaneous.

- The left-to-right dimension indicates the temporal order of procedures in the experiment (sometimes indicated with an arrow).

- The symbol R indicates random assignment.

- Separation of parallel rows by a horizontal line indicates that comparison groups are not equal (or equated) by random assignment. No horizontal line between the groups displays random assignment of individuals to treatment groups.

In the following examples, this notation is used to illustrate pre-experimental, quasi-experimental, true experimental, and single-subject designs.

Example 8.2   *Pre-Experimental Designs*

One-Shot Case Study

This design involves an exposure of a group to a treatment followed by a measure.

Group A X——————————O

One-Group Pre-Test–Post-Test Design

This design includes a pre-test measure followed by a treatment and a post-test for a single group.

Group A 01————X————02

Static Group Comparison or Post-Test-Only With Nonequivalent Groups

Experimenters use this design after implementing a treatment. After the treatment, the researcher selects a comparison group and provides a post-test to both the experimental group(s) and the comparison group(s).

Group A X——————————O
_____
Group B ——————————O

Alternative Treatment Post-Test-Only With Nonequivalent Groups Design

This design uses the same procedure as the Static Group Comparison, with the exception that the nonequivalent comparison group received a different treatment.

Group A X1——————————O
_____
Group B X2——————————O

Example 8.3   *Quasi-Experimental Designs*

Nonequivalent (Pre-Test and Post-Test) Control-Group Design

In this design, a popular approach to quasi-experiments, the experimental group A and the control group B are selected without random assignment. Both groups take a pre-test and post-test. Only the experimental group receives the treatment.

Group A O————X————O
————————————————
Group B O————————O

Single-Group Interrupted Time-Series Design

In this design, the researcher records measures for a single group both before and after a treatment.

Group A O—O—O—O—X—O—O—O—O

Control-Group Interrupted Time-Series Design

A modification of the Single-Group Interrupted Time-Series design in which two groups of participants, not randomly assigned, are observed over time. A treatment is administered to only one of the groups (i.e., Group A).

Group A O—O—O—O—X—O—O—O—O
————————————————————
Group B O—O—O—O—O—O—O—O—O

Example 8.4   *True Experimental Designs*

Pre-Test–Post-Test Control-Group Design

A traditional, classical design, this procedure involves random assignment of participants to two groups. Both groups are administered both a pre-test and a post-test, but the treatment is provided only to experimental Group A.

Group A R————O————X————O
Group B R————O————————O

Post-Test-Only Control-Group Design

This design controls for any confounding effects of a pre-test and is a popular experimental design. The participants are randomly assigned to groups, a treatment is given only to the experimental group, and both groups are measured on the post-test.

Group A R————————X————O
Group B R————————————O

Solomon Four-Group Design

A special case of a 2 X 2 factorial design, this procedure involves the random assignment of participants to four groups. Pre-tests and treatments are varied for the four groups. All groups receive a post-test.

Group A R————O————X————O
Group B R————O————————O
Group C R————————X————O
Group D R————————————O

Example 8.5   *Single-Subject Designs*

A-B-A Single-Subject Design

This design involves multiple observations of a single individual. The target behavior of a single individual is established over time and is referred to as a baseline behavior. The baseline behavior is assessed, the treatment provided, and then the treatment is withdrawn.

Baseline A   Treatment B Baseline A

O–O–O–O–O–X–X–X–X–X–O–O–O–O–O–O

Threats to Validity

There are several threats to validity that will raise questions about an experimenter's ability to conclude that the intervention affects an outcome and not some other factor. Experimental researchers need to identify potential threats to the internal validity of their experiments and design them so that these threats will not likely arise or are minimized. There are two types of threats to validity: internal threats and external threats. **Internal validity threats** are experimental procedures, treatments, or experiences of the participants that threaten the researcher's ability to draw correct inferences from the data about the population in an experiment. Table 8.5 displays these threats, provides a description of each one of them, and suggests potential responses by the researcher so that the threat may not occur. There are those involving participants (i.e., history, maturation, regression, selection, and mortality), those related to the use of an experimental treatment that the researcher manipulates (i.e., diffusion, compensatory and resentful demoralization, and compensatory rivalry), and those involving procedures used in the experiment (i.e., testing and instruments).

Potential threats to external validity also must be identified and designs created to minimize these threats. **External validity threats** arise when experimenters draw incorrect inferences from the sample data to other persons, other settings, and past or future situations. As shown in Table 8.6, these threats arise because of the characteristics of individuals selected for the sample, the uniqueness of the setting, and the timing of the experiment. For example, threats to external validity arise when the researcher generalizes beyond the groups in the experiment to other racial or social groups not under study, to settings not studied, or to past or future situations. Steps for addressing these potential issues are also presented in Table 8.6.

Other threats that might be mentioned in the method section are the threats to **statistical conclusion validity** that arise when experimenters draw inaccurate inferences from the data because of inadequate statistical

Table 8.5 Types of Threats to Internal Validity

Type of Threat to Internal Validity	Description of Threat	In Response, Actions the Researcher Can Take
History	Because time passes during an experiment, events can occur that unduly influence the outcome beyond the experimental treatment.	The researcher can have both the experimental and control groups experience the same external events.
Maturation	Participants in an experiment may mature or change during the experiment, thus influencing the results.	The researcher can select participants who mature or change at the same rate (e.g., same age) during the experiment.
Regression	Participants with extreme scores are selected for the experiment. Naturally, their scores will probably change during the experiment. Scores, over time, regress toward the mean.	A researcher can select participants who do not have extreme scores as entering characteristics for the experiment.
Selection	Participants can be selected who have certain characteristics that predispose them to have certain outcomes (e.g., they are brighter).	The researcher can select participants randomly so that characteristics have the probability of being equally distributed among the experimental groups.
Mortality	Participants drop out during an experiment due to many possible reasons. The outcomes are thus unknown for these individuals.	A researcher can recruit a large sample to account for dropouts or compare those who drop out with those who continue, in terms of the outcome.
Diffusion of treatment	Participants in the control and experimental groups communicate with each other. This communication can influence how both groups score on the outcomes.	The researcher can keep the two groups as separate as possible during the experiment.

(Continued)

Type of Threat to Internal Validity	Description of Threat	In Response, Actions the Researcher Can Take
Table 8.5 (Continued)		
Compensatory/resentful demoralization	The benefits of an experiment may be unequal or resented when only the experimental group receives the treatment (e.g., experimental group receives therapy and the control group receives nothing).	The researcher can provide benefits to both groups, such as giving the control group the treatment *after* the experiment ends or giving the control group some different type of treatment *during* the experiment.
Compensatory rivalry	Participants in the control group feel that they are being devalued, as compared to the experimental group, because they do not experience the treatment.	The researcher can take steps to create equality between the two groups, such as reducing the expectations of the control group.
Testing	Participants become familiar with the outcome measure and remember responses for later testing.	The researcher can have a longer time interval between administrations of the outcome or use different items on a later test than were used in an earlier test.
Instrumentation	The instrument changes between a pre-test and post-test, thus impacting the scores on the outcome.	The researcher can use the same instrument for the pre-test and post-test measures.

SOURCE: Adapted from Creswell (2008).

power or the violation of statistical assumptions. Threats to **construct validity** occur when investigators use inadequate definitions and measures of variables.

Practical **research tips** for proposal writers to address validity issues are as follows:

● Identify the potential threats to validity that may arise in your study. A separate section in a proposal may be composed to advance this threat.

● Define the exact type of threat and what potential issue it presents to your study.

Table 8.6 Types of Threats to External Validity		
Types of Threats to External Validity	Description of Threat	In Response, Actions the Researcher Can Take
Interaction of selection and treatment	Because of the narrow characteristics of participants in the experiment, the researcher cannot generalize to individuals who do not have the characteristics of participants.	The researcher restricts claims about groups to which the results cannot be generalized. The researcher conducts additional experiments with groups with different characteristics.
Interaction of setting and treatment	Because of the characteristics of the setting of participants in an experiment, a researcher cannot generalize to individuals in other settings.	The researcher needs to conduct additional experiments in new settings to see if the same results occur as in the initial setting.
Interaction of history and treatment	Because results of an experiment are time-bound, a researcher cannot generalize the results to past or future situations.	The researcher needs to replicate the study at later times to determine if the same results occur as in the earlier time.

SOURCE: Adapted from Creswell (2008).

● Discuss how you plan to address the threat in the design of your experiment.

● Cite references to books that discuss the issue of threats to validity, such as Cook and Campbell (1979); Creswell (2008); Reichardt and Mark (1998); Shadish, Cook, & Campbell (2001); and Tuckman (1999).

The Procedure

A proposal developer needs to describe in detail the procedure for conducting the experiment. A reader should be able to understand the design being used, the observations, the treatment, and the timeline of activities.

● Discuss a step-by-step approach for the procedure in the experiment. For example, Borg and Gall (1989, p. 679) outlined six steps typically used in the procedure for a pre-test–post-test control group design with matching participants in the experimental and control groups:

1. Administer measures of the dependent variable or a variable closely correlated with the dependent variable to the research participants.

2. Assign participants to matched pairs on the basis of their scores on the measures described in Step 1.

3. Randomly assign one member of each pair to the experimental group and the other member to the control group.

4. Expose the experimental group to the experimental treatment and administer no treatment or an alternative treatment to the control group.

5. Administer measures of the dependent variables to the experimental and control groups.

6. Compare the performance of the experimental and control groups on the post-test(s) using tests of statistical significance.

Data Analysis

Tell the reader about the types of statistical analysis that will be used during the experiment.

● Report the descriptive statistics calculated for observations and measures at the pre-test or post-test stage of experimental designs. These statistics are means, standard deviations, and ranges.

● Indicate the inferential statistical tests used to examine the hypotheses in the study. For experimental designs with categorical information (groups) on the independent variable and continuous information on the dependent variable, researchers use t tests or univariate analysis of variance (ANOVA), analysis of covariance (ANCOVA), or multivariate analysis of variance (MANOVA—multiple dependent measures). (Several of these tests are mentioned in Table 8.3, presented earlier.) In factorial designs, both interaction and main effects of ANOVA are used. When data on a pretest or post-test show marked deviation from a normal distribution, use nonparametric statistical tests.

● For single-subject research designs, use line graphs for baseline and treatment observations for abscissa (horizontal axis) units of time and the ordinate (vertical axis) target behavior. Each data point is plotted separately on the graph, and the data points are connected by lines (e.g., see Neuman & McCormick, 1995). Occasionally, tests of statistical significance, such as t tests, are used to compare the pooled mean of the baseline and the treatment phases, although such procedures may violate the assumption of independent measures (Borg & Gall, 1989).

● With increasing frequency, experimental researchers report both statistical results of hypothesis testing and confidence intervals and effect size as indicators of practical significance of the findings. A **confidence interval** is an interval estimate of the range of upper and lower statistical values that are consistent with the observed data and are likely to contain the actual population mean. An **effect size** identifies the strength of the conclusions about group differences or the relationships among variables in quantitative studies. The calculation of effect size varies for different statistical tests.

Interpreting Results

The final step in an experiment is to interpret the findings in light of the hypotheses or research questions set forth in the beginning. In this interpretation, address whether the hypotheses or questions were supported or whether they were refuted. Consider whether the treatment that was implemented actually made a difference for the participants who experienced them. Suggest why or why not the results were significant, drawing on past literature that you reviewed (Chapter 2), the theory used in the study (Chapter 3), or persuasive logic that might explain the results. Address whether the results might have occurred because of inadequate experimental procedures, such as threats to internal validity, and indicate how the results might be generalized to certain people, settings, and times. Finally, indicate the implications of the results for the population studied or for future research.

Example 8.6   *An Experimental Method Section*

The following is a selected passage from a quasi-experimental study by Enns and Hackett (1990) that demonstrates many of the components in an experimental design. Their study addressed the general issue of matching client and counselor interests along the dimensions of attitudes toward feminism. They hypothesized that feminist participants would be more receptive to a radical feminist counselor than would nonfeminist participants and that nonfeminist participants would be more receptive to a nonsexist and liberal feminist counselor. Except for a limited discussion about data analysis and an interpretation section found in the discussion of their article, their approach contains the elements of a good method section for an experimental study.

Method

Participants

The participants were 150 undergraduate women enrolled in both lower- and upper-division courses in sociology, psychology, and communications at a midsized university and a community college, both on the west coast. (*The authors described the participants in this study.*)

(Continued)

(Continued)

Design and Experimental Manipulation

This study used a 3 X 2 X 2 factorial design: Orientation of Counselor (non-sexist-humanistic, liberal feminist, or radical feminist) X Statement of Values (implicit or explicit) X Participants' Identification with Feminism (feminist or nonfeminist). Occasional missing data on particular items were handled by a pairwise deletion procedure. *(Authors identified the overall design.)*

The three counseling conditions, nonsexist-humanistic, liberal, and radical femi-nist, were depicted by 10 min videotape vignettes of a second counseling ses-sion between a female counselor and a female client. . . . The implicit statement of values condition used the sample interview only; the counselor's values were therefore implicit in her responses. The explicit statement of values condition was created by adding to each of the three counseling conditions a 2-min leader that portrayed the counselor describing to the client her counseling approach and associated values including for the two feminist conditions a description of her feminist philosophical orientation, liberal or radical. . . . Three counseling scripts were initially developed on the basis of distinctions between nonsexist-humanistic, liberal, and radical feminist philosophies and attendant counseling implications. Client statements and the outcome of each interview were held constant, whereas counselor responses differed by approach. *(Authors described the three treatment conditions variables manipulated in the study.)*

Instruments

Manipulation checks. As a check on participants' perception of the experi-mental manipulation and as an assessment of participants' perceived similarity to the three counselors, two subscales of Berryman-Fink and Verderber's (1985) Attributions of the Term Feminist Scale were revised and used in this study as the Counselor Description Questionnaire (CDQ) and the Personal Description Questionnaire (PDQ). . . . Berryman-Fink and Verderber (1985) reported internal consistency reliabilities of .86 and .89 for the original versions of these two subscales. *(Authors discussed the instruments and the reliability of the scales for the dependent variable in the study.)*

Procedure

All experimental sessions were conducted individually. The experimenter, an advanced doctoral student in counseling psychology, greeted each subject, explained the purpose of the study as assessing students' reactions to coun-seling, and administered the ATF. The ATF was then collected and scored while each subject completed a demographic data form and reviewed a set of instructions for viewing the videotape. The first half of the sample was randomly assigned to one of the twelve videotapes (3 Approaches X 2 Statements X 2 Counselors), and a median was obtained on the ATF. The median for the first half of the sample was then used to categorize the second half of the group as feminist or nonfeminist, and the remainder of the participants was randomly assigned to conditions separately from each feminist orientation group to ensure nearly equal cell sizes. The median on the final sample was checked and a few participants recategorized by the final median split, which resulted in 12 or 13 participants per cell.

After viewing the videotape that corresponded to their experimental assignment, participants completed the dependent measures and were debriefed. (pp. 35–36; *Authors described the procedure used in the experiment.*)

SUMMARY

This chapter identified essential components in designing a method procedure for a survey or experimental study. The outline of steps for a survey study began with a discussion about the purpose, the identification of the population and sample, the survey instruments to be used, the relationship between the variables, the research questions, specific items on the survey, and steps to be taken in the analysis and the interpretation of the data from the survey. In the design of an experiment, the researcher identifies participants in the study, the variables—the treatment conditions and the outcome variables—and the instruments used for pre-tests and post-tests and the materials to be used in the treatments. The design also includes the specific type of experiment, such as a pre-experimental, quasi-experimental, true experiment, or single-subject design. Then the researcher draws a figure to illustrate the design, using appropriate notation. This is followed by comments about potential threats to internal and external validity (and possibly statistical and construct validity) that relate to the experiment, the statistical analysis used to test the hypotheses or research questions, and the interpretation of the results.

Writing Exercises

1. Design a plan for the procedures to be used in a survey study. Review the checklist in Table 8.1 after you write the section to determine if all components have been addressed.

2. Design a plan for procedures for an experimental study. Refer to Table 8.4 after you complete your plan to determine if all questions have been addressed adequately.

WRITING EXERCISES

ADDITIONAL READINGS

Babbie, E. (1990). *Survey research methods* (2nd ed.). Belmont, CA: Wadsworth.

Earl Babbie provides a thorough, detailed text about all aspects of survey design. He reviews the types of designs, the logic of sampling, and examples of designs. He also discusses the conceptualization of a survey instrument and its scales. He then provides useful ideas about administering a questionnaire and processing the results. Also included is a discussion about data analysis with attention to constructing and understanding tables and writing a survey report. This book is detailed, informative, and technically oriented toward students at the intermediate or advanced level of survey research.

Campbell, D. T., & Stanley, J. C. (1963). Experimental and quasi-experimental designs for research. In N. L. Gage (Ed.), *Handbook of research on teaching* (pp. 1–76). Chicago: Rand-McNally.

This chapter in the Gage *Handbook* is the classical statement about experimental designs. Campbell and Stanley designed a notation system for experiments that is still used today; they also advanced the types of experimental designs, beginning with factors that jeopardize internal and external validity, the pre-experimental design types, true experiments, quasi-experimental designs, and correlational and ex post facto designs. The chapter presents an excellent summary of types of designs, their threats to validity, and statistical procedures to test the designs. This is an essential chapter for students beginning their study of experimental studies.

Fink, A. (2002). *The survey kit* (2nd ed.). Thousand Oaks, CA: Sage.

"The Survey Kit," is composed of multiple books and edited by Arlene Fink. An overview of the books in this series is provided in the first volume. As an introduction to the volumes, Fink discusses all aspects of survey research, including how to ask questions, how to conduct surveys, how to engage in telephone interviews, how to sample, and how to measure validity and reliability. Much of the discussion is oriented toward the beginning survey researcher, and the numerous examples and excellent illustrations make it a useful tool to learn the basics of survey research.

Fowler, F. J. (2002). *Survey research methods.* (3rd ed.). Thousand Oaks, CA: Sage.

Floyd Fowler provides a useful text about the decisions that go into the design of a survey research project. He addresses use of alternative sampling procedures, ways of reducing nonresponse rates, data collection, design of good questions, employing sound interviewing techniques, preparation of surveys for analysis, and ethical issues in survey designs.

Keppel, G. (1991). *Design and analysis: A researcher's handbook* (3rd ed.). Englewood Cliffs, NJ: Prentice-Hall.

Geoffrey Keppel provides a detailed, thorough treatment of the design of experiments from the principles of design to the statistical analysis of experimental data. Overall, this book is for the mid-level to advanced statistics student who seeks to understand the design and statistical analysis of experiments. The introductory chapter presents an informative overview of the components of experimental designs.

Lipsey, M. W. (1990). *Design sensitivity: Statistical power for experimental research.* Newbury Park, CA: Sage.

Mark Lipsey has authored a major book on the topics of experimental designs and statistical power of those designs. Its basic premise is that an experiment needs to have sufficient sensitivity to detect those effects it purports to investigate. The book explores statistical power and includes a table to help researchers identify the appropriate size of groups in an experiment.

Neuman, S. B., & McCormick, S. (Eds.). (1995). *Single-subject experimental research: Applications for literacy.* Newark, DE: International Reading Association.

Susan Neuman and Sandra McCormick have edited a useful, practical guide to the design of single-subject research. They present many examples of different types of designs, such as reversal designs and multiple-baseline designs, and they enumerate the statistical procedures that might be involved in analyzing the single-subject data. One chapter, for example, illustrates the conventions for displaying data on line graphs. Although this book cites many applications in literacy, it has broad application in the social and human sciences.

CHAPTER NINE

Qualitative Procedures

Qualitative procedures demonstrate a different approach to scholarly inquiry than methods of quantitative research. Qualitative inquiry employs different philosophical assumptions; strategies of inquiry; and methods of data collection, analysis, and interpretation. Although the processes are similar, qualitative procedures rely on text and image data, have unique steps in data analysis, and draw on diverse strategies of inquiry.

In fact, the strategies of inquiry chosen in a qualitative project have a dramatic influence on the procedures, which, even within strategies, are anything but uniform. Looking over the landscape of qualitative procedures shows diverse perspectives ranging from social justice thinking (Denzin & Lincoln, 2005), to ideological perspectives (Lather, 1991), to philosophical stances (Schwandt, 2000), to systematic procedural guidelines (Creswell, 2007; Corbin & Strauss, 2007). All perspectives vie for center stage in this unfolding model of inquiry called qualitative research.

This chapter attempts to combine many perspectives, provide general procedures, and use examples liberally to illustrate variations in strategies. This discussion draws on thoughts provided by several authors writing about qualitative proposal design (e.g., see Berg, 2001; Marshall & Rossman, 2006; Maxwell, 2005; Rossman & Rallis, 1998). The topics in a proposal section on procedures are characteristics of qualitative research, the research strategy, the role of the researcher, steps in data collection and analysis, strategies for validity, the accuracy of findings, and narrative structure. Table 9.1 shows a checklist of questions for designing qualitative procedures.

THE CHARACTERISTICS OF QUALITATIVE RESEARCH

For many years, proposal writers had to discuss the characteristics of qualitative research and convince faculty and audiences as to their legitimacy. Now these discussions are less frequently found in the literature and there is some consensus as to what constitutes qualitative inquiry. Thus, my suggestions about this section of a proposal are as follows:

Table 9.1	A Checklist of Questions for Designing a Qualitative Procedure
_____	Are the basic characteristics of qualitative studies mentioned?
_____	Is the specific type of qualitative strategy of inquiry to be used in the study mentioned? Is the history of, a definition of, and applications for the strategy mentioned?
_____	Does the reader gain an understanding of the researcher's role in the study (past historical, social, cultural experiences, personal connections to sites and people, steps in gaining entry, and sensitive ethical issues)?
_____	Is the purposeful sampling strategy for sites and individuals identified?
_____	Are the specific forms of data collection mentioned and a rationale given for their use?
_____	Are the procedures for recording information during the data collection procedure mentioned (such as protocols)?
_____	Are the data analysis steps identified?
_____	Is there evidence that the researcher has organized the data for analysis?
_____	Has the researcher reviewed the data generally to obtain a sense of the information?
_____	Has coding been used with the data?
_____	Have the codes been developed to form a description or to identify themes?
_____	Are the themes interrelated to show a higher level of analysis and abstraction?
_____	Are the ways that the data will be represented mentioned—such as in tables, graphs, and figures?
_____	Have the bases for interpreting the analysis been specified (personal experiences, the literature, questions, action agenda)?
_____	Has the researcher mentioned the outcome of the study (developed a theory, provided a complex picture of themes)?
_____	Have multiple strategies been cited for validating the findings?

- Review the needs of potential audiences for the proposal. Decide whether audience members are knowledgeable enough about the characteristics of qualitative research that this section is not necessary.

- If there is some question about their knowledge, present the basic characteristics of qualitative research in the proposal and possibly discuss a recent qualitative research journal article (or study) to use as an example to illustrate the characteristics.

● Several lists of characteristics might be used (e.g., Bogdan & Biklen, 1992; Eisner, 1991; Hatch, 2002; LeCompte & Schensul, 1999; Marshall & Rossman, 2006), but I will rely on a composite analysis of several of these writers that I incorporated into my book on qualitative inquiry (Creswell, 2007). My list captures both traditional perspectives and the newer advocacy, participatory, and self-reflexive perspectives of qualitative inquiry. Here are the characteristics of qualitative research, presented in no specific order of importance:

- Natural setting—Qualitative researchers tend to collect data in the field at the site where participants experience the issue or problem under study. They do not bring individuals into a lab (a contrived situation), nor do they typically send out instruments for individuals to complete. This up close information gathered by actually talking directly to people and seeing them behave and act within their context is a major characteristic of qualitative research. In the natural setting, the researchers have face-to-face interaction over time.

- Researcher as key instrument—Qualitative researchers collect data themselves through examining documents, observing behavior, or interviewing participants. They may use a protocol—an instrument for collecting data—but the researchers are the ones who actually gather the information. They do not tend to use or rely on questionnaires or instruments developed by other researchers.

- Multiple sources of data—Qualitative researchers typically gather multiple forms of data, such as interviews, observations, and documents, rather than rely on a single data source. Then the researchers review all of the data, make sense of it, and organize it into categories or themes that cut across all of the data sources.

- Inductive data analysis—Qualitative researchers build their patterns, categories, and themes from the bottom up, by organizing the data into increasingly more abstract units of information. This inductive process illustrates working back and forth between the themes and the database until the researchers have established a comprehensive set of themes. It may also involve collaborating with the participants interactively, so that participants have a chance to shape the themes or abstractions that emerge from the process.

- Participants' meanings—In the entire qualitative research process, the researcher keeps a focus on learning the meaning that the participants hold about the problem or issue, not the meaning that the researchers bring to the research or writers express in the literature.

- Emergent design—The research process for qualitative researchers is emergent. This means that the initial plan for research cannot be tightly prescribed, and all phases of the process may change or shift after the researcher enters the field and begins to collect data. For example, the questions may change, the forms of data collection may shift, and the individuals studied and the sites visited may be modified. The key idea behind qualitative research is to learn about the problem or issue from participants and to address the research to obtain that information.

- Theoretical lens—Qualitative researchers often use lens to view their studies, such as the concept of culture, central to ethnography, or gendered, racial, or class differences from the theoretical orientations discussed in Chapter 3. Sometimes the study may be organized around identifying the social, political, or historical context of the problem under study.

- Interpretive—Qualitative research is a form of interpretive inquiry in which researchers make an interpretation of what they see, hear, and understand. Their interpretations cannot be separated from their own backgrounds, history, contexts, and prior understandings. After a research report is issued, the readers make an interpretation as well as the participants, offering yet other interpretations of the study. With the readers, the participants, and the researchers all making interpretations, it is apparent how multiple views of the problem can emerge.

- Holistic account—Qualitative researchers try to develop a complex picture of the problem or issue under study. This involves reporting multiple perspectives, identifying the many factors involved in a situation, and generally sketching the larger picture that emerges. A visual model of many facets of a process or a central phenomenon aid in establishing this holistic picture (see, for example, Creswell & Brown, 1992).

STRATEGIES OF INQUIRY

Beyond these general characteristics are more specific strategies of inquiry. These strategies focus on data collection, analysis, and writing, but they originate out of disciplines and flow throughout the process of research (e.g., types of problems, ethical issues of importance; Creswell, 2007b). Many strategies exist, such as the 28 approaches identified by Tesch (1990), the 19 types in Wolcott's (2001) tree, and the 5 approaches to qualitative inquiry by Creswell (2007). As discussed in Chapter 1, I recommend that qualitative researchers choose from among the possibilities, such as narrative, phenomenology, ethnography, case study, and grounded theory. I selected these five because they are popular across the social and health sciences today. Others exist that have been addressed adequately in qualitative books, such as participatory action research (Kemmis & Wilkinson, 1998) or discourse analysis (Cheek, 2004). For the five approaches, researchers might study individuals (narrative, phenomenology); explore processes, activities, and events (case study, grounded theory); or learn about broad culture-sharing behavior of individuals or groups (ethnography).

In writing a procedure for a qualitative proposal, consider the following **research tips:**

- Identify the specific approach to inquiry that you will be using.

- Provide some background information about the strategy, such as its discipline origin, the applications of it, and a brief definition of it (see Chapter 1 for the five strategies of inquiry).

- Discuss why it is an appropriate strategy to use in the proposed study.

- Identify how the use of the strategy will shape the types of questions asked (see Morse, 1994, for questions that relate to strategies), the form of data collection, the steps of data analysis, and the final narrative.

THE RESEARCHER'S ROLE

As mentioned in the list of characteristics, qualitative research is interpretative research, with the inquirer typically involved in a sustained and intensive experience with participants. This introduces a range of strategic, ethical, and personal issues into the qualitative research process (Locke et al., 2007). With these concerns in mind, inquirers explicitly identify reflexively their biases, values, and personal background, such as gender, history, culture, and socioeconomic status, that may shape their interpretations formed during a study. In addition, gaining entry to a research site and the ethical issues that might arise are also elements of the researcher's role.

- Include statements about past experiences that provide background data through which the audience can better understand the topic, the setting, or the participants and the researcher's interpretation of the phenomenon.

- Comment on connections between the researcher and the participants and on the research sites. "Backyard" research (Glesne & Peshkin, 1992) involves studying the researcher's own organization, or friends, or immediate work setting. This often leads to compromises in the researcher's ability to disclose information and raises difficult power issues. Although data collection may be convenient and easy, the problems of reporting data that are biased, incomplete, or compromised are legion. If studying the backyard is necessary, employ multiple strategies of validity (as discussed later) to create reader confidence in the accuracy of the findings.

- Indicate steps taken to obtain permission from the Institutional Review Board (see Chapter 4) to protect the rights of human participants. Attach, as an appendix, the approval letter from the IRB and discuss the process involved in securing permission.

● Discuss steps taken to gain entry to the setting and to secure permission to study the participants or situation (Marshall & Rossman, 2006). It is important to gain access to research or archival sites by seeking the approval of **gatekeepers**, individuals at the research site that provide access to the site and allow or permit the research to be done. A brief proposal might need to be developed and submitted for review by gatekeepers. Bogdan and Biklen (1992) advance topics that could be addressed in such a proposal:

- Why was the site chosen for study?
- What activities will occur at the site during the research study?
- Will the study be disruptive?
- How will the results be reported?
- What will the gatekeeper gain from the study?

● Comment about sensitive ethical issues that may arise (see Chapter 3, and Berg, 2001). For each issue raised, discuss how the research study will address it. For example, when studying a sensitive topic, it is necessary to mask names of people, places, and activities. In this situation, the process for masking information requires discussion in the proposal.

DATA COLLECTION PROCEDURES

Comments about the role of the researcher set the stage for discussion of issues involved in collecting data. The data collection steps include setting the boundaries for the study, collecting information through unstructured or semistructured observations and interviews, documents, and visual materials, as well as establishing the protocol for recording information.

● Identify the *purposefully selected* sites or individuals for the proposed study. The idea behind qualitative research is to **purposefully select** participants or sites (or documents or visual material) that will best help the researcher understand the problem and the research question. This does not necessarily suggest random sampling or selection of a large number of participants and sites, as typically found in *quantitative* research. A discussion about participants and site might include four aspects identified by Miles and Huberman (1994): the *setting* (where the research will take place), the *actors* (who will be observed or interviewed), the *events* (what the actors will be observed or interviewed doing), and the *process* (the evolving nature of events undertaken by the actors within the setting).

● Indicate the type or types of data to be collected. In many qualitative studies, inquirers collect multiple forms of data and spend a considerable time in the natural setting gathering information. The collection procedures in qualitative research involve four basic types, as shown in Table 9.2.

Table 9.2	Qualitative Data Collection Types, Options, Advantages, and Limitations		
Data Collection Types	**Options Within Types**	**Advantages of the Type**	**Limitations of the Type**
Observations	• Complete participant—researcher conceals role • Observer as participant—role of researcher is known • Participant as observer—observation role secondary to participant role • Complete observer—researcher observes without participating	• Researcher has a first-hand experience with participant. • Researcher can record information as it occurs. • Unusual aspects can be noticed during observation. • Useful in exploring topics that may be uncomfortable for participants to discuss.	• Researcher may be seen as intrusive. • Private information may be observed that researcher cannot report. • Researcher may not have good attending and observing skills. • Certain participants (e.g., children) may present special problems in gaining rapport.
Interviews	• Face-to-face—one-on-one, in-person interview • Telephone—researcher interviews by phone • Focus group—researcher interviews participants in a group • E-mail internet interview	• Useful when participants cannot be directly observed. • Participants can provide historical information. • Allows researcher control over the line of questioning.	• Provides indirect information filtered through the views of interviewees. • Provides information in a designated place rather than the natural field setting. • Researcher's presence may bias responses. • Not all people are equally articulate and perceptive.

(Continued)

Table 9.2 (Continued)			
Data Collection Types	Options Within Types	Advantages of the Type	Limitations of the Type
Documents	• Public documents, such as minutes of meetings, or newspapers • Private documents, such as journals, diaries, or letters	• Enables a researcher to obtain the language and words of participants. • Can be accessed at a time convenient to researcher—an unobtrusive source of information. • Represents data which are thoughtful in that participants have given attention to compiling them. • As written evidence, it saves a researcher the time and expense of transcribing.	• Not all people are equally articulate and perceptive. • May be protected information unavailable to public or private access. • Requires the researcher to search out the information in hard-to-find places. • Requires transcribing or optically scanning for computer entry. • Materials may be incomplete. • The documents may not be authentic or accurate.
Audio-Visual Materials	• Photographs • Videotapes • Art objects • Computer software • Film	• May be an unobtrusive method of collecting data. • Provides an opportunity for participants to directly share their reality. • It is creative in that it captures attention visually.	• May be difficult to interpret. • May not be accessible publicly or privately. • The presence of an observer (e.g., photographer) may be disruptive and affect responses.

NOTE: This table includes material taken from Merriam (1998), Bogdan & Biklen (1992), and Creswell (2007).

- **Qualitative observations** are those in which the researcher takes field notes on the behavior and activities of individuals at the research site. In these field notes, the researcher records, in an unstructured or semistructured way (using some prior questions that the inquirer wants to know), activities at the research site. Qualitative observers may also engage in roles varying from a nonparticipant to a complete participant.

- In **qualitative interviews,** the researcher conducts face-to-face interviews with participants, interviews participants by telephone, or engages in focus group interviews, with six to eight interviewees in each group. These interviews involve unstructured and generally open-ended questions that are few in number and intended to elicit views and opinions from the participants.

- During the process of research, the investigator may collect **qualitative documents.** These may be public documents (e.g., newspapers, minutes of meetings, official reports) or private documents (e.g., personal journals and diaries, letters, e-mails).

- A final category of qualitative data consists of **qualitative audio and visual materials.** This data may take the form of photographs, art objects, videotapes, or any forms of sound.

● In a discussion about data collection forms, be specific about the types and include arguments concerning the strengths and weaknesses of each type, as discussed in Table 9.2.

● Include data collection types that go beyond typical observations and interviews. These unusual forms create reader interest in a proposal and can capture useful information that observations and interviews may miss. For example, examine the compendium of types of data in Table 9.3 that can be used, to stretch the imagination about possibilities, such as gathering sounds or tastes, or using cherished items to elicit comments during an interview.

DATA RECORDING PROCEDURES

Before entering the field, qualitative researchers plan their approach to data recording. The proposal should identify what data the researcher will record and the procedures for recording data.

● Use a *protocol* for recording observational data. Researchers often engage in multiple observations during the course of a qualitative study and use an **observational protocol** for recording information while observing. This may be a single page with a dividing line down the middle to separate *descriptive notes* (portraits of the participants, a reconstruction of dialogue, a description of the physical setting, accounts of particular

Table 9.3   A List of Qualitative Data Collection Approaches

Observations

- Gather field notes by conducting an observation as a participant.
- Gather field notes by conducting an observation as an observer.
- Gather field notes by spending more time as a participant than as an observer.
- Gather field notes by spending more time as an observer than as a participant.
- Gather field notes first by observing as an outsider and then moving into the setting and observing as an insider.

Interviews

- Conduct an unstructured, open-ended interview and take interview notes.
- Conduct an unstructured, open-ended interview, audiotape the interview, and transcribe it.
- Conduct a semistructured interview, audiotape the interview, and transcribe the interview.
- Conduct a focus group interview, audiotape the interview, and transcribe it.
- Conduct different types of interviews: email, face-to-face, focus group, online focus group, telephone interviews

Documents

- Keep a journal during the research study.
- Have a participant keep a journal or diary during the research study.
- Collect personal letters from participants.
- Analyze public documents (e.g., official memos, minutes, records, archival material).
- Examine autobiographies and biographies.
- Have participants take photographs or videotapes (i.e., photo elicitation).
- Chart audits
- Medical records

Audio-visual Materials

- Examine physical trace evidence (e.g., footprints in the snow).
- Videotape or film a social situation or an individual or group.
- Examine photographs or videotapes.
- Collect sounds (e.g., musical sounds, a child's laughter, car horns honking).
- Collect e-mail messages.
- Collect cell phone text messages.
- Examine possessions or ritual objects.
- Collect sounds, smells, tastes, or any stimuli of the senses.

SOURCE: Adapted from Creswell (2007).

events, or activities) from *reflective notes* (the researcher's personal thoughts, such as "speculation, feelings, problems, ideas, hunches, impressions, and prejudices" Bogdan & Biklen, 1992, p. 121). Also written on this form might be *demographic information* about the time, place, and date of the field setting where the observation takes place.

● Use an **interview protocol** for asking questions and recording answers during a qualitative interview. This protocol includes the following components:

- A heading (date, place, interviewer, interviewee)

- Instructions for the interviewer to follow so that standard procedures are used from one interview to another

- The questions (typically an ice-breaker question at the beginning followed by 4–5 questions that are often the subquestions in a qualitative research plan, followed by some concluding statement or a question, such as, "Who should I visit with to learn more about my questions?"

- Probes for the 4–5 questions, to follow up and ask individuals to explain their ideas in more detail or to elaborate on what they have said

- Space between the questions to record responses

- A final thank-you statement to acknowledge the time the interviewee spent during the interview (see Creswell, 2007)

● Researchers record information from interviews by making handwritten notes, by audiotaping, or by videotaping. Even if an interview is taped, I recommend that researchers take notes, in the event that recording equipment fails. If audiotaping is used, researchers need to plan in advance for the transcription of the tape.

● The recording of documents and visual materials can be based on the researcher's structure for taking notes. Typically, notes reflect information about the document or other material as well as key ideas in the documents. It is helpful to note whether the information represents primary material (i.e., information directly from the people or situation under study) or secondary material (i.e., secondhand accounts of the people or situation written by others). It is also helpful to comment on the reliability and value of the data source.

DATA ANALYSIS AND INTERPRETATION

Discussion of the plan for analyzing the data might have several components. The process of data analysis involves making sense out of text and image data. It involves preparing the data for analysis, conducting different analyses, moving deeper and deeper into understanding the data (some qualitative researchers like to think of this as peeling back the layers of an onion), representing the data, and making an interpretation of the larger meaning of the data. Several generic processes might be stated in the proposal that convey a sense of the overall activities of qualitative data analysis, such as the following drawn from my own thoughts (Creswell, 2007) and those of Rossman and Rallis (1998):

● It is an ongoing process involving continual reflection about the data, asking analytic questions, and writing memos throughout the study. I say that qualitative data analysis is conducted concurrently with gathering data, making interpretations, and writing reports. While interviews are going on, for example, the researcher may be analyzing an interview collected earlier, writing memos that may ultimately be included as a narrative in the final report, and organizing the structure of the final report.

● Data analysis involves collecting open-ended data, based on asking general questions and developing an analysis from the information supplied by participants.

● Often we see qualitative data analysis reported in journal articles and books that is a generic form of analysis. In this approach, the researcher collects qualitative data, analyzes it for themes or perspectives, and reports 4–5 themes. I consider this approach to be basic qualitative analysis; today many qualitative researchers go beyond this generic analysis to add a procedure within one of the qualitative strategies of inquiry. For example, *grounded theory* has systematic steps (Corbin & Strauss, 2007; Strauss & Corbin, 1990, 1998). These involve generating categories of information (open coding), selecting one of the categories and positioning it within a theoretical model (axial coding), and then explicating a story from the interconnection of these categories (selective coding). *Case study* and *ethnographic research* involve a detailed description of the setting or individuals, followed by analysis of the data for themes or issues (see Stake, 1995; Wolcott, 1994). *Phenomenological research* uses the analysis of significant statements, the generation of meaning units, and the development of what Moustakas (1994) calls an essence description. *Narrative research* employs restorying the participants' stories using structural devices, such as plot, setting, activities, climax, and denouement (Clandinin & Connelly, 2000). As these examples illustrate, the processes as well as the terms differ from one analytic strategy to another.

● Despite these analytic differences depending on the type of strategy used, qualitative inquirers often use a general procedure and convey in the proposal the steps in data analysis. An ideal situation is to blend the general steps with the specific research strategy steps. An overview of the data analysis process is seen in Figure 9.1. As a **research tip,** I urge researchers to look at qualitative data analysis as following steps from the specific to the general and as involving multiple levels of analysis.

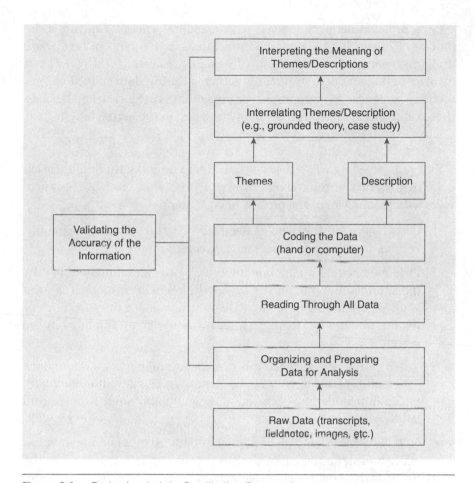

Figure 9.1   Data Analysis in Qualitative Research

This figure suggests a linear, hierarchical approach building from the bottom to the top, but I see it as more interactive in practice; the various stages are interrelated and not always visited in the order presented. These levels are emphasized in the following steps:

Step 1. *Organize and prepare* the data for analysis. This involves transcribing interviews, optically scanning material, typing up field notes, or sorting and arranging the data into different types depending on the sources of information.

Step 2. Read through all the data. A first step is to obtain a *general sense* of the information and to reflect on its overall meaning. What general ideas are participants saying? What is the tone of the ideas? What is the impression of the overall depth, credibility, and use of the information? Sometimes qualitative researchers write notes in margins or start recording general thoughts about the data at this stage.

Step 3. Begin detailed analysis with a coding process. **Coding** is the process of organizing the material into chunks or segments of text before bringing meaning to information (Rossman & Rallis, 1998, p. 171). It involves taking text data or pictures gathered during data collection, segmenting sentences (or paragraphs) or images into categories, and labeling those categories with a term, often a term based in the actual language of the participant (called an *in vivo* term).

Before proceeding to Step 4, consider some remarks that will provide detailed guidance for the coding process. Tesch (1990, pp. 142–145) provides a useful analysis of the process in eight steps:

1. Get a sense of the whole. Read all the transcriptions carefully. Perhaps jot down some ideas as they come to mind.

2. Pick one document (i.e., one interview)—the most interesting one, the shortest, the one on the top of the pile. Go through it, asking yourself, "What is this about?" Do not think about the substance of the information but its underlying meaning. Write thoughts in the margin.

3. When you have completed this task for several participants, make a list of all topics. Cluster together similar topics. Form these topics into columns, perhaps arrayed as major topics, unique topics, and leftovers.

4. Now take this list and go back to your data. Abbreviate the topics as codes and write the codes next to the appropriate segments of the text. Try this preliminary organizing scheme to see if new categories and codes emerge.

5. Find the most descriptive wording for your topics and turn them into categories. Look for ways of reducing your total list of categories by grouping topics that relate to each other. Perhaps draw lines between your categories to show interrelationships.

6. Make a final decision on the abbreviation for each category and alphabetize these codes.

7. Assemble the data material belonging to each category in one place and perform a preliminary analysis.

8. If necessary, recode your existing data.

These eight steps engage a researcher in a systematic process of analyzing textual data. Variations exist in this process. As a **research tip,** I encourage qualitative researchers to analyze their data for material that can address the following:

● Codes on topics that readers would expect to find, based on the past literature and common sense

- Codes that are surprising and that were not anticipated at the beginning of the study

- Codes that are unusual, and that are, in and of themselves, of conceptual interest to readers (e.g., in Asmussen and Creswell, 1995, we identified *retriggering* as one of the codes/themes in the analysis that suggested a new dimension for us to a gunman incident on campus and that seemed to connect with experiences of others on campus)

- Codes that address a larger theoretical perspective in the research

As an alternative conceptualization, consider the list by Bogdan and Biklen (1992, pp. 166–172) of the types of codes that they look for in a qualitative database:

- Setting and context codes

- Perspectives held by subjects

- Subjects' ways of thinking about people and objects

- Process codes

- Activity codes

- Strategy codes

- Relationship and social structure codes

- Preassigned coding schemes

One further issue about coding is whether the researcher should (a) develop codes *only* on the basis of the emerging information collected from participants, (b) use predetermined codes and then fit the data to them, or (c) use some combination of predetermined and emerging codes. The traditional approach in the social sciences is to allow the codes to emerge during the data analysis. In the health sciences, a popular approach is to use predetermined codes based on the theory being examined. In this case, the researchers might develop a **qualitative codebook**, a table or record that contains a list of predetermined codes that researchers use for coding the data. This codebook might be composed with the names of codes in one column, a definition of codes in another column, and then specific instances (e.g., line numbers) in which the code was found in the transcripts. Having such a codebook is invaluable when multiple researchers are coding the data from different transcripts. This codebook can evolve and change during a study based on close analysis of the data, even when the researcher is not starting from an emerging code perspective. For researchers who have a distinct theory they want to test in their projects, I would recommend that a preliminary codebook be developed for coding the data and permit the codebook to develop and change based on the information learned during the data analysis. The use of a codebook is especially helpful for fields in which *quantitative* research dominates and a more systematic approach to *qualitative* research is needed.

Returning to the general coding process, some researchers have found it useful to hand code qualitative transcripts or information, sometimes using color code schemes and to cut and paste text segments onto note cards. This is a laborious and time-consuming approach. Others tend to use qualitative computer software programs to help code, organize, and sort information that will be useful in writing the qualitative study. Several excellent computer software programs are available, and they have similar features: good tutorials and demonstration CDs, ability to incorporate both text and image (e.g., photographs) data, the feature of storing and organizing data, the search capacity of locating all text associated with specific codes, interrelated codes for making queries of the relationship among codes, and the import and export of qualitative data to *quantitative* programs, such as spreadsheets or data analysis programs.

The basic idea behind these programs is that using the computer is an efficient means for storing and locating qualitative data. Although the researcher still needs to go through each line of text (as in transcriptions) and assign codes, this process may be faster and more efficient than hand coding. Also, in large databases, the researcher can quickly locate all passages (or text segments) coded the same and determine whether participants are responding to a code idea in similar or different ways. Beyond this, the computer program can facilitate comparing different codes (e.g., How do males and females—the first code of gender—differ in terms of their attitudes to smoking—a second code) These are just a few features of the software programs that make them a logical choice for qualitative data analysis over hand coding. As with any software program, qualitative software programs require time and skill to learn and employ effectively, although books for learning the programs are widely available (e.g., Weitzman & Miles, 1995).

Most of the programs are available only on the PC platform. The computer software programs that my staff and I use in my research office are these:

● MAXqda (http://www.maxqda.com/). This is an excellent PC-based program from Germany that helps researchers systematically evaluate and interpret qualitative texts. It has all of the features mentioned earlier.

● Atlas.ti (http://www.atlasti.com). This is another PC-based program from Germany that enables a researcher to organize text, graphic, audio, and visual data files, along with coding, memos and findings, into a project.

● QSR NVivo (http://www.qsrinternational.com/) This program, from Australia, features the popular software program N6 (or Nud.ist) and NVivo concept mapping in combination. It is available only for Windows PC.

● HyperRESEARCH (http://www.researchware.com/). This is a program available for either the MAC or PC. It is an easy-to-use qualitative software package enabling users to code, retrieve, build theories, and conduct analyses of the data.

Step 4. Use the coding process to generate a description of the setting or people as well as categories or themes for analysis. *Description* involves a detailed rendering of information about people, places, or events in a setting. Researchers can generate codes for this description. This analysis is useful in designing detailed descriptions for case studies, ethnographies, and narrative research projects. Then use the coding to generate a small number of *themes* or categories, perhaps five to seven categories for a research study. These themes are the ones that appear as major findings in qualitative studies and are often used to create headings in the findings sections of studies. They should display multiple perspectives from individuals and be supported by diverse quotations and specific evidence.

Beyond identifying the themes during the coding process, qualitative researchers can do much with themes to build additional layers of complex analysis. For example, researchers interconnect themes into a story line (as in narratives) or develop them into a theoretical model (as in grounded theory). Themes are analyzed for each individual case and across different cases (as in case studies) or shaped into a general description (as in phenomenology). Sophisticated qualitative studies go beyond description and theme identification and into complex theme connections.

Step 5. Advance how the description and themes will be *represented* in the qualitative narrative. The most popular approach is to use a narrative passage to convey the findings of the analysis. This might be a discussion that mentions a chronology of events, the detailed discussion of several themes (complete with subthemes, specific illustrations, multiple perspectives from individuals, and quotations) or a discussion with interconnecting themes. Many qualitative researchers also use visuals, figures, or tables as adjuncts to the discussions. They present a process model (as in grounded theory), advance a drawing of the specific research site (as in ethnography), or convey descriptive information about each participant in a table (as in case studies and ethnographies).

Step 6. A final step in data analysis involves making an **interpretation** or meaning of the data. Asking, "What were the lessons learned?" captures the essence of this idea (Lincoln & Guba, 1985). These *lessons* could be the researcher's personal interpretation, couched in the understanding that the inquirer brings to the study from her or his own culture, history, and experiences. It could also be a meaning derived from a comparison of the findings with information gleaned from the *literature* or *theories*. In this way, authors suggest that the findings confirm past information or diverge from it. It can also suggest *new questions* that need to be asked—questions raised by the data and analysis that the inquirer had not foreseen earlier in the study. One way ethnographers can end a study, says Wolcott (1994), is to ask further questions. The questioning approach is also used in advocacy and participatory approaches to qualitative research. Moreover, when qualitative researchers use a theoretical lens, they can form interpretations that call for action agendas for reform and change. Thus, interpretation in qualitative research can take many forms, be adapted for different types of designs, and be flexible to convey personal, research-based, and action meanings.

RELIABILITY, VALIDITY, AND GENERALIZABILITY

Although validation of findings occurs throughout the steps in the process of research (as shown in Figure 9.1), this discussion focuses on it to enable a researcher to write a passage into a proposal on the procedures for validating the findings that will be undertaken in a study. Proposal developers need to convey the steps they will take in their studies to check for the accuracy and credibility of their findings.

Validity does not carry the same connotations in qualitative research as it does in quantitative research, nor is it a companion of reliability (examining stability or consistency of responses) or generalizability (the external validity of applying results to new settings, people, or samples; both are discussed in Chapter 8). **Qualitative validity** means that the researcher checks for the accuracy of the findings by employing certain procedures, while **qualitative reliability** indicates that the researcher's approach is consistent across different researchers and different projects (Gibbs, 2007).

How do qualitative researchers check to determine if their approaches are consistent or reliable? Yin (2003) suggests that qualitative researchers need to document the procedures of their case studies and to document as many of the steps of the procedures as possible. He also recommends setting up a detailed case study protocol and database. Gibbs (2007) suggests several **reliability procedures:**

● Check transcripts to make sure that they do not contain obvious mistakes made during transcription.

● Make sure that there is not a drift in the definition of codes, a shift in the meaning of the codes during the process of coding. This can be accomplished by constantly comparing data with the codes and by writing memos about the codes and their definitions (see the discussion on a qualitative codebook).

● For team research, coordinate the communication among the coders by regular documented meetings and by sharing the analysis.

● Cross-check codes developed by different researchers by comparing results that are independently derived.

Proposal writers need to include several of these procedures as evidence that they will have consistent results in their proposed study. I recommend that several procedures be mentioned in a proposal and that single researchers find another person who can cross-check their codes, for what I call **intercoder agreement** (or cross-checking). Such an agreement might be based on whether two or more coders agree on codes used for the same passages in the text (it is not that they code the same passage of text, but whether another coder would code it with the same or a similar code). Statistical procedures or reliability subprograms in qualitative computer software packages can then be used to determine the level of consistency of coding. Miles and Huberman (1994) recommend that the consistency of the coding be in agreement at least 80% of the time for good qualitative reliability.

Validity, on the other hand, is one of the strengths of qualitative research, and it is based on determining whether the findings are accurate from the standpoint of the researcher, the participant, or the readers of an account (Creswell & Miller, 2000). Terms abound in the qualitative literature that speak to this idea, such as *trustworthiness, authenticity*, and *credibility* (Creswell & Miller, 2000), and it is a much-discussed topic (Lincoln & Guba, 2000).

A procedural perspective that I recommend for research proposals is to identify and discuss one or more strategies available to check the accuracy of the findings. The researcher actively incorporates **validity strategies** into their proposal. I recommend the use of multiple strategies, and these should enhance the researcher's ability to assess the accuracy of findings as well as convince readers of that accuracy. There are eight primary strategies, organized from those most frequently used and easy to implement to those occasionally used and more difficult to implement:

● *Triangulate* different data sources of information by examining evidence from the sources and using it to build a coherent justification for themes. If themes are established based on converging several sources of data or perspectives from participants, then this process can be claimed as adding to the validity of the study.

● Use *member checking* to determine the accuracy of the qualitative findings through taking the final report or specific descriptions or themes back to participants and determining whether these participants feel that they are accurate. This does not mean taking back the raw transcripts to check for accuracy; instead, the researcher takes back parts of the polished product, such as the themes, the case analysis, the grounded theory, the cultural description, and so forth. This procedure can involve conducting a follow-up interview with participants in the study and providing an opportunity for them to comment on the findings.

● Use *rich, thick description* to convey the findings. This description may transport readers to the setting and give the discussion an element of shared experiences. When qualitative researchers provide detailed descriptions of the setting, for example, or provide many perspectives about a theme, the results become more realistic and richer. This procedure can add to the validity of the findings.

● Clarify the *bias* the researcher brings to the study. This self-reflection creates an open and honest narrative that will resonate well with readers. Reflectivity has been mentioned as a core characteristic of qualitative research. Good qualitative research contains comments by the researchers about how their interpretation of the findings is shaped by their background, such as their gender, culture, history, and socioeconomic origin.

● Also present *negative* or *discrepant information* that runs counter to the themes. Because real life is composed of different perspectives that do not always coalesce, discussing contrary information adds to the credibility of an account. A researcher can accomplish this in discussing evidence about a theme. Most evidence will build a case for the theme; researchers can also present information that contradicts the general perspective of the theme. By presenting this contradictory evidence, the account becomes more realistic and hence valid.

● Spend *prolonged time* in the field. In this way, the researcher develops an in-depth understanding of the phenomenon under study and can convey detail about the site and the people that lends credibility to the narrative account. The more experience that a researcher has with participants in their actual setting, the more accurate or valid will be the findings.

● Use *peer debriefing* to enhance the accuracy of the account. This process involves locating a person (a peer debriefer) who reviews and asks questions about the qualitative study so that the account will resonate with people other than the researcher. This strategy—involving an interpretation beyond the researcher and invested in another person—adds validity to an account.

● Use an *external auditor* to review the entire project. As distinct from a peer debriefer, this auditor is not familiar with the researcher or the project and can provide an objective assessment of the project throughout the process of research or at the conclusion of the study. The role is similar to that of a fiscal auditor, and specific questions exist that auditors might ask (Lincoln & Guba, 1985). The procedure of having an independent investigator look over many aspects of the project (e.g., accuracy of transcription, the relationship between the research questions and the data, the level of data analysis from the raw data through interpretation) enhances the overall validity of a qualitative study.

Qualitative generalization is a term that is used in a limited way in qualitative research, since the intent of this form of inquiry is not to generalize findings to individuals, sites, or places outside of those under study (see Gibbs, 2007, for his cautionary note about qualitative generalizability). In fact, the value of qualitative research lies in the particular description and themes developed *in context* of a specific site. *Particularity* rather than *generalizability* (Greene & Caracelli, 1997) is the hallmark of qualitative research. However, there are a few discussions in the qualitative literature about generalizability, especially as applied to case study research in which the inquirer studies several cases. Yin (2003), for example, feels that qualitative case study results can be generalized to some *broader theory*. The generalization occurs when qualitative researchers study additional cases and generalize findings to the new cases. It is the same as the *replication logic* used in experimental research. However, to repeat a case study's findings in a new case setting requires good documentation of qualitative procedures, such as a protocol for documenting the problem in detail and the development of a thorough case study database (Yin, 2003).

THE QUALITATIVE WRITE-UP

A plan for a qualitative procedure should end with some comments about the narrative that emerges from the data analysis. Numerous varieties of narratives exist, and examples from scholarly journals illustrate models. In a plan for a study, consider advancing several points about the narrative.

The basic procedure in reporting the results of a qualitative study are to develop descriptions and themes from the data (see Figure 9.1), to present these descriptions and themes that convey multiple perspectives from participants and detailed descriptions of the setting or individuals. Using a qualitative strategy of inquiry, these results may also provide a chronological narrative of an individual's life (narrative research), a detailed description of their experiences (phenomenology), a theory generated from the data (grounded theory), a detailed portrait of a culture-sharing group (ethnography), or an in-depth analysis of one or more cases (case study).

Given these different strategies, the findings and interpretation sections of a plan for a study might discuss how the sections will be presented: as objective accounts, fieldwork experiences (Van Maanen, 1988), a chronology, a process model, an extended story, an analysis by cases or across cases, or a detailed descriptive portrait (Creswell, 2007).

At the specific level, some **writing strategies** might be as follows:

● Use quotes and vary their length from short to long embedded passages.

● Script conversation and report the conversation in different languages to reflect cultural sensitivity.

- Present text information in tabular form (e.g., matrices, comparison tables of different codes).

- Use the wording from participants to form codes and theme labels.

- Intertwine quotations with (the author's) interpretations.

- Use indents or other special formatting of the manuscript to call attention to quotations from participants.

- Use the first person "I" or collective "we" in the narrative form.

- Use metaphors and analogies (see, for example, Richardson, 1990, who discusses some of these forms).

- Use the narrative approach typically used within a qualitative strategy of inquiry (e.g., description in case studies and ethnographies, a detailed story in narrative research).

- Describe how the narrative outcome will be compared with theories and the general literature on the topic. In many qualitative articles, researchers discuss the literature at the end of the study (see the discussion in Chapter 2).

Example 9.1 *Qualitative Procedures*

The following is an example of a qualitative procedure written as part of a doctoral proposal (Miller, 1992). Miller's project was an ethnographic study of first-year experiences of the president of a 4-year college. As I present this discussion, I refer back to the sections addressed in this chapter and highlight them in boldfaced type. Also, I have maintained Miller's use of the term *informant*, although today, the more appropriate term, *participant*, should be used.

The Qualitative Research Paradigm

The qualitative research paradigm has its roots in cultural anthropology and American sociology (Kirk & Miller, 1986). It has only recently been adopted by educational researchers (Borg & Gall, 1989). The intent of qualitative research is to understand a particular social situation, event, role, group, or interaction (Locke, Spirduso, & Silverman, 1987). It is largely an investigative process where the researcher gradually makes sense of a social phenomenon by contrasting, comparing, replicating, cataloguing and classifying the object of study (Miles & Huberman, 1984). Marshall and Rossman (1989) suggest that this entails immersion in the everyday life of the setting chosen for the study; the researcher enters the informants' world and through ongoing interaction, seeks the informants' perspectives and meanings. *(Qualitative assumptions are mentioned.)*

Scholars contend that qualitative research can be distinguished from quantitative methodology by numerous unique characteristics that are inherent in the design. The following is a synthesis of commonly articulated assumptions regarding characteristics presented by various researchers.

1. Qualitative research occurs in natural settings, where human behavior and events occur.

2. Qualitative research is based on assumptions that are very different from quantitative designs. Theory or hypotheses are not established a priori.

3. The researcher is the primary instrument in data collection rather than some inanimate mechanism (Eisner, 1991; Frankel & Wallen, 1990; Lincoln & Guba, 1985; Merriam, 1988).

4. The data that emerge from a qualitative study are descriptive. That is, data are reported in words (primarily the participant's words) or pictures, rather than in numbers (Fraenkel & Wallen, 1990; Locke et al., 1987; Marshall & Rossman, 1989; Merriam, 1988).

5. The focus of qualitative research is on participants' perceptions and experiences, and the way they make sense of their lives (Fraenkel & Wallen, 1990; Locke et al., 1987; Merriam, 1988). The attempt is therefore to understand not one, but multiple realities (Lincoln & Guba, 1985).

6. Qualitative research focuses on the process that is occurring as well as the product or outcome. Researchers are particularly interested in understanding how things occur (Fraenkel & Wallen, 1990; Merriam, 1988).

7. Idiographic interpretation is utilized. In other words, attention is paid to particulars; and data is interpreted in regard to the particulars of a case rather than generalizations.

8. Qualitative research is an emergent design in its negotiated outcomes. Meanings and interpretations are negotiated with human data sources because it is the subjects' realities that the researcher attempts to reconstruct (Lincoln & Guba, 1985; Merriam, 1988).

9. This research tradition relies on the utilization of tacit knowledge (intuitive and felt knowledge) because often the nuances of the multiple realities can be appreciated most in this way (Lincoln & Guba, 1985). Therefore, data are not quantifiable in the traditional sense of the word.

(Continued)

(Continued)

10. Objectivity and truthfulness are critical to both research traditions. However, the criteria for judging a qualitative study differ from quantitative research. First and foremost, the researcher seeks believability, based on coherence, insight and instrumental utility (Eisner, 1991) and trustworthiness (Lincoln & Guba, 1985) through a process of verification rather than through traditional validity and reliability measures. *(Qualitative characteristics are mentioned.)*

The Ethnographic Research Design

This study will utilize the ethnographic research tradition. This design emerged from the field of anthropology, primarily from the contributions of Bronislaw Malinowski, Robert Park and Franz Boas (Jacob, 1987; Kirk & Miller, 1986). The intent of ethnographic research is to obtain a holistic picture of the subject of study with emphasis on portraying the everyday experiences of individuals by observing and interviewing them and relevant others (Fraenkel & Wallen, 1990). The ethnographic study includes in-depth interviewing and continual and ongoing participant observation of a situation (Jacob, 1987) and in attempting to capture the whole picture reveals how people describe and structure their world (Fraenkel & Wallen, 1990). *(The author used the ethnographic approach.)*

The Researcher's Role

Particularly in qualitative research, the role of the researcher as the primary data collection instrument necessitates the identification of personal values, assumptions and biases at the outset of the study. The investigator's contribution to the research setting can be useful and positive rather than detrimental (Locke et al., 1987). My perceptions of higher education and the college presidency have been shaped by my personal experiences. From August 1980 to May 1990 I served as a college administrator on private campuses of 600 to 5,000. Most recently (1987–1990), I served as the Dean for Student Life at a small college in the Midwest. As a member of the President's cabinet, I was involved with all top level administrative cabinet activities and decisions and worked closely with the faculty, cabinet officers, president and board of trustees. In addition to reporting to the president, I worked with him through his first year in office. I believe this understanding of the context and role enhances my awareness, knowledge and sensitivity to many of the challenges, decisions and issues encountered as a first year president and will assist me in working with the informant in this study. I bring knowledge of both the structure of higher education and of the role the college presidency. Particular attention will be paid to the role of the new president in initiating change, relationship building, decision making, and providing leadership and vision.

Due to previous experiences working closely with a new college president, I bring certain biases to this study. Although every effort will be made to ensure objectivity, these biases may shape the way I view and understand the data I collect and the way I interpret my experiences. I commence this study with the perspective that the college presidency is a diverse and often difficult position. Though expectations are immense, I question how much power the president has to initiate change and provide leadership and vision. I view the first year as critical; filled with adjustments, frustrations, unanticipated surprises and challenges. *(Author reflected on her role in the study.)*

Bounding the Study

Setting

This study will be conducted on the campus of a state college in the Midwest. The college is situated in a rural Midwestern community. The institution's 1,700 students nearly triple the town's population of 1,000 when classes are in session. The institution awards associate, bachelor and master's degrees in 51 majors.

Actors

The informant in this study is the new President of a state college in the Midwest. The primary informant in this study is the President, However, I will be observing him in the context of administrative cabinet meetings. The president's cabinet includes three Vice Presidents (Academic Affairs, Administration, Student Affairs) and two Deans (Graduate Studies and Continuing Education).

Events

Using ethnographic research methodology, the focus of this study will be the everyday experiences and events of the new college president, and the perceptions and meaning attached to those experiences as expressed by the informant. This includes the assimilation of surprising events or information, and making sense of critical events and issues that arise.

Processes

Particular attention will be paid to the role of the new president in initiating change, relationship building, decision making, and providing leadership and vision. *(Author mentioned data collection boundaries.)*

Ethical Considerations

Most authors who discuss qualitative research design address the importance of ethical considerations (Locke et al., 1982; Marshall & Rossman, 1989; Merriam, 1988; Spradley, 1980). First and foremost, the researcher has

(Continued)

(Continued)

an obligation to respect the rights, needs, values, and desires of the informant(s). To an extent, ethnographic research is always obtrusive. Participant observation invades the life of the informant (Spradley, 1980) and sensitive information is frequently revealed. This is of particular concern in this study where the informant's position and institution are highly visible. The following safeguards will be employed to protect the informant's rights: 1) the research objectives will be articulated verbally and in writing so that they are clearly understood by the informant (including a description of how data will be used), 2) written permission to proceed with the study as articulated will be received from the informant, 3) a research exemption form will be filed with the Institutional Review Board (Appendixes B1 and B2), 4) the informant will be informed of all data collection devices and activities, 5) verbatim transcriptions and written interpretations and reports will be made available to the informant, 6) the informant's rights, interests and wishes will be considered first when choices are made regarding reporting the data, and 7) the final decision regarding informant anonymity will rest with the informant. *(Author addressed ethical issues and IRB review.)*

Data Collection Strategies

Data will be collected from February through May, 1992. This will include a minimum of bi-monthly, 45 minute recorded interviews with the informant (initial interview questions, Appendix C), bimonthly two hour observations of administrative cabinet meetings, bi-monthly two hour observations of daily activities and bi-monthly analysis of the president's calendar and documents (meeting minutes, memos, publications). In addition, the informant has agreed to record impressions of his experiences, thoughts and feelings in a taped diary (guidelines for recorded reflection, Appendix D). Two follow-up interviews will be scheduled for the end of May 1992 (See Appendix E for proposed timeline and activity schedule). *(The author proposed to use face-to-face interviews, participate as observer, and obtain private documents.)*

To assist in the data collection phase I will utilize a field log, providing a detailed account of ways I plan to spend my time when I am on-site, and in the transcription and analysis phase (also comparing this record to how time is actually spent). I intend to record details related to my observations in a field notebook and keep a field diary to chronicle my own thinking, feeling, experiences and perceptions throughout the research process. *(The author recorded descriptive and reflective information.)*

Data Analysis Procedures

Merriam (1988) and Marshall and Rossman (1989) contend that data collection and data analysis must be a simultaneous process in qualitative research. Schatzman and Strauss (1973) claim that qualitative data analysis primarily entails classifying things, persons, and events and the properties which characterize them. Typically throughout the data analysis process ethnographers index or code their data using as many categories as possible (Jacob, 1987). They seek to identify and describe patterns and themes from the perspective of the participant(s), then attempt to understand and explain these patterns and themes (Agar, 1980). During data analysis the data will be organized categorically and chronologically, reviewed repeatedly, and continually coded. A list of major ideas that surface will be chronicled (as suggested by Merriam, 1988). Taped interviews and the participant's taped diary will be transcribed verbatim. Field notes and diary entries will be regularly reviewed. *(Author described steps in data analysis.)*

In addition, the data analysis process will be aided by the use of a qualitative data analysis computer program called HyperQual. Raymond Padilla (Arizona State University) designed HyperQual in 1987 for use with the Macintosh computer. HyperQual utilizes HyperCard software and facilitates the recording and analysis of textual and graphic data. Special stacks are designated to hold and organize data. Using HyperQual the researcher can directly "enter field data, including interview data, observations, researcher's memos, and illustrations...(and) tag (or code) all or part of the source data so that chunks of data can be pulled out and then be reassembled in a new and illuminating configuration" (Padilla, 1989, pp. 69–70). Meaningful data chunks can be identified, retrieved, isolated, grouped and regrouped for analysis. Categories or code names can be entered initially or at a later date. Codes can be added, changed or deleted with HyperQual editor and text can be searched for key categories, themes, words or phrases. *(Author mentions the proposed use of computer software for data analysis.)*

Verification

In ensuring internal validity, the following strategies will be employed:

1. Triangulation of data—Data will be collected through multiple sources to include interviews, observations and document analysis;

2. Member checking—The informant will serve as a check throughout the analysis process. An ongoing dialogue regarding my interpretations of the informant's reality and meanings will ensure the truth value of the data;

3. Long terms and repeated observations at the research site—Regular and repeated observations of similar phenomena and settings will occur on-site over a four month period of time;

(Continued)

(Continued)

4. Peer examination—a doctoral student and graduate assistant in the Educational Psychology Department will serve as a peer examiner;

5. Participatory modes of research—The informant will be involved in most phases of this study, from the design of the project to checking interpretations and conclusions; and

6. Clarification of researcher bias—At the outset of this study researcher bias will be articulated in writing in the dissertation proposal under the heading, "The Researcher's Role."

The primary strategy utilized in this project to ensure external validity will be the provision of rich, thick, detailed descriptions so that anyone interested in transferability will have a solid framework for comparison (Merriam, 1988). Three techniques to ensure reliability will be employed in this study. First, the researcher will provide a detailed account of the focus of the study, the researcher's role, the informant's position and basis for selection, and the context from which data will be gathered (LeCompte & Goetz, 1984). Second, triangulation or multiple methods of data collection and analysis will be used, which strengthens reliability as well as internal validity (Merriam, 1988). Finally, data collection and analysis strategies will be reported in detail in order to provide a clear and accurate picture of the methods used in this study. All phases of this project will be subject to scrutiny by an external auditor who is experienced in qualitative research methods. *(Author identified strategies of validity to be used in the study.)*

Reporting the Findings

Lofland (1974) suggests that although data collection and analysis strategies are similar across qualitative methods, the way the findings are reported is diverse. Miles and Huberman (1984) address the importance of creating a data display and suggest that narrative text has been the most frequent form of display for qualitative data. This is a naturalistic study. Therefore, the results will be presented in descriptive, narrative form rather than as a scientific report. Thick description will be the vehicle for communicating a holistic picture of the experiences of a new college president. The final project will be a construction of the informant's experiences and the meanings he attaches to them. This will allow readers to vicariously experience the challenges he encounters and provide a lens through which readers can view the subject's world. *(Outcomes of the study were mentioned.)*

SUMMARY

This chapter explored the steps that go into developing and writing a qualitative procedure. Recognizing the variation that exists in qualitative studies, the chapter advances a general guideline for procedures. This guideline includes a discussion about the general characteristics of qualitative research if audiences are not familiar with this approach to research. These characteristics are that the research takes place in the natural setting, relies on the researcher as the instrument for data collection, employs multiple methods of data collection, is inductive, is based on participants' meanings, is emergent, often involves the use of a theoretical lens, is interpretive, and is holistic. The guideline recommends mentioning a strategy of inquiry, such as the study of individuals (narrative, phenomenology), the exploration of processes, activities and events (case study, grounded theory), or the examination of broad culture-sharing behavior of individuals or groups (ethnography). The choice of strategy needs to be presented and defended. Further, the proposal needs to address the role of the researcher: past experiences, personal connections to the site, steps to gain entry, and sensitive ethical issues. Discussion of data collection should include the purposeful sampling approach and the forms of data to be collected (i.e., observations, interviews, documents, audiovisual materials). It is useful to also indicate the types of data recording protocols that will be used. Data analysis is an ongoing process during research. It involves analyzing participant information, and researchers typically employ general analysis steps as well as those steps found within a specific strategy of inquiry. More general steps include organizing and preparing the data, an initial reading through the information, coding the data, developing from the codes a description and thematic analysis, using computer programs, representing the findings in tables, graphs, and figures, and interpreting the findings. These interpretations involve stating lessons learned, comparing the findings with past literature and theory, raising questions, and/or advancing an agenda for reform. The proposal should also contain a section on the expected outcomes for the study. Finally, an additional important step in planning a proposal is to mention the strategies that will be used to validate the accuracy of the findings, demonstrate the reliability of procedures, and discuss the role of generalizability.

Writing Exercises

1. Write a plan for the procedure to be used in your qualitative study. After writing the plan, use Table 9.1 as a checklist to determine the comprehensiveness of your plan.

2. Develop a table that lists, in a column on the left, the steps you plan to take to analyze your data. In a column on the right, indicate the steps as they apply directly to your project, the research strategy you plan to use, and data that you have collected.

ADDITIONAL READINGS

Marshall, C., & Rossman, G. B. (2006). *Designing qualitative research* (4th ed.). Thousand Oaks, CA: Sage.

Catherine Marshall and Gretchen Rossman introduce the procedures for designing a qualitative study and a qualitative proposal. The topics covered are comprehensive. They include building a conceptual framework around a study; the logic and assumptions of the overall design and methods; methods of data collection and procedures for managing, recording, and analyzing qualitative data; and the resources needed for a study, such as time, personnel, and funding. This is a comprehensive and insightful text from which both beginners and more experienced qualitative researchers can learn.

Flick, U. (Ed.). (2007). *The Sage Qualitative Research Kit.* London: Sage.

This is an eight-volume kit edited by Uwe Flick that is authored by different world-class qualitative researchers and was created to collectively address the core issues that arise when researchers actually do qualitative research. It addresses how to plan and design a qualitative study, the collection and production of qualitative data, the analysis of qualitative data (e.g., visual data, discourse analysis), and the issues of quality in qualitative research. Overall, it presents a recent, up-to-date window into the field of qualitative research.

Creswell, J. W. (2007). *Qualitative inquiry and research design: Choosing among five approaches* (2nd ed.). Thousand Oaks, CA: Sage.

Sometimes those who write about qualitative research take a philosophical stance toward the topic and readers are left without an understanding of the procedures and practices actually used in designing and conducting a qualitative study. My book takes five approaches to qualitative inquiry—narrative research, phenomenology, grounded theory, ethnography, and case study—and discusses how the procedures for conducting these forms of inquiry are both similar and different. In the end, readers can more easily choose which of the five would best suit their research problems as well as their personal styles of research.

CHAPTER TEN

Mixed Methods Procedures

With the development and perceived legitimacy of both qualitative and quantitative research in the social and human sciences, mixed methods research, employing the combination of quantitative and qualitative approaches, has gained popularity. This popularity is because research methodology continues to evolve and develop, and mixed methods is another step forward, utilizing the strengths of both qualitative and quantitative research. Also, the problems addressed by social and health science researchers are complex, and the use of either quantitative or qualitative approaches by themselves is inadequate to address this complexity. The interdisciplinary nature of research, as well, contributes to the formation of research teams with individuals with diverse methodological interests and approaches. Finally, there is more insight to be gained from the combination of both qualitative and quantitative research than either form by itself. Their combined use provides an expanded understanding of research problems.

This chapter brings together many of the threads introduced in the earlier chapters: It extends the discussion about the philosophical assumptions of a pragmatic philosophy, the combined use of qualitative and quantitative modes of inquiry, and the use of multiple methods introduced in Chapter 1. It also extends the discussion about research problems that incorporate the need both to explore and explain (Chapter 5). It follows a purpose statement and research questions focused on understanding a problem using both qualitative and quantitative methods (Chapters 6 and 7), and it advances the reasons for using multiple forms of data collection and analysis (Chapters 8 and 9).

COMPONENTS OF MIXED METHODS PROCEDURES

Mixed methods research has evolved a set of procedures that proposal developers can use in planning a mixed methods study. In 2003, the *Handbook of*

Mixed Methods in the Social & Behavior Sciences (Tashakkori & Teddlie, 2003) was published, providing the first comprehensive overview of this strategy of inquiry. Now several journals emphasize mixed methods research, such as the *Journal of Mixed Methods Research, Quality and Quantity*, and *Field Methods*, while numerous others actively encourage this form of inquiry (e.g., *International Journal of Social Research Methodology, Qualitative Health Research, Annals of Family Medicine*). Numerous published research studies have incorporated mixed methods research in the social and human sciences in diverse fields such as occupational therapy (Lysack & Krefting, 1994), interpersonal communication (Boneva, Kraut, & Frohlich, 2001), AIDS prevention (Janz et al., 1996), dementia caregiving (Weitzman & Levkoff, 2000), mental health (Rogers, Day, Randall, & Bentall, 2003), and in middle-school science (Houtz, 1995). New books arrive each year solely devoted to mixed methods research (Bryman, 2006; Creswell & Plano Clark, 2007; Greene, 2007; Plano Clark & Creswell, 2008; Tashakkori & Teddlie, 1998).

A checklist of questions for designing a mixed methods study appears in Table 10.1. These components call for advancing the nature of mixed methods research and the type of strategy being proposed for the study. They also include the need for a visual model of this approach, the specific procedures of data collection and analysis, the researcher's role, and the structure for presenting the final report. Following the discussion of each of these components, an example of a procedures section from a mixed methods study is presented to show how to apply these ideas.

THE NATURE OF MIXED METHODS RESEARCH

Because mixed methods research is relatively new in the social and human sciences as a distinct research approach, it is useful to convey a basic definition and description of the approach in a proposal. This might include the following:

● Trace a brief history of its evolution. Several sources identify its inception in psychology and in the multitrait–multimethod matrix of Campbell and Fiske (1959) to interest in converging or triangulating different quantitative and qualitative data sources (Jick, 1979) and on to the development of a distinct methodology of inquiry (see Creswell & Plano Clark, 2007; Tashakkori & Teddlie, 1998).

● Define mixed methods research by incorporating the definition in Chapter 1 that focuses on combining both quantitative and qualitative research and methods in a research study (see a more expanded view of defining mixed methods research in Johnson, Onwuegbuzie, & Turner, 2007). Highlight the reasons why researchers employ a mixed methods design (e.g., to broaden understanding by incorporating both qualitative

Table 10.1	A Checklist of Questions for Designing a Mixed Methods Procedure
_____	Is a basic definition of mixed methods research provided?
_____	Is a reason given for using both quantitative and qualitative approaches (or data)?
_____	Does the reader have a sense for the potential use of a mixed methods design?
_____	Are the criteria identified for choosing a mixed methods strategy?
_____	Is the strategy identified, and are its criteria for selection given?
_____	Is a visual model presented that illustrates the research strategy?
_____	Is the proper notation used in presenting the visual model?
_____	Are procedures of data collection and analysis mentioned as they relate to the model?
_____	Are the sampling strategies for both quantitative and qualitative data collection mentioned? Do they relate to the strategy?
_____	Are specific data analysis procedures indicated? Do they relate to the strategy?
_____	Are the procedures for validating both the quantitative and qualitative data discussed?
_____	Is the narrative structure mentioned, and does it relate to the type of mixed methods strategy being used?

and quantitative research, or to use one approach to better understand, explain, or build on the results from the other approach). Also note that the mixing of the two might be within one study or among several studies in a program of inquiry. Recognize that many different terms are used for this approach, such as *integrating, synthesis, quantitative and qualitative methods, multimethod,* and *mixed methodology,* but that recent writings use the term *mixed methods* (Bryman, 2006; Tashakkori & Teddlie, 2003).

● Briefly discuss the growth of interest in mixed methods research as expressed in books, journal articles, diverse disciplines, and funded projects (see Creswell & Plano Clark, 2007 for a discussion about the many initiatives that contribute to mixed methods today).

● Note the challenges this form of research poses for the inquirer. These include the need for extensive data collection, the time-intensive nature of analyzing both text and numeric data, and the requirement for the researcher to be familiar with both quantitative and qualitative forms of research.

TYPES OF MIXED METHODS STRATEGIES AND VISUAL MODELS

There have been several typologies for classifying and identifying types of mixed methods strategies that proposal developers might use in their proposed mixed methods study. Creswell and Plano Clark (2007) identify 12 classification systems drawn from the fields of evaluation, nursing, public health, education policy and research, and social and behavioral research. In these classifications, authors use diverse terms for their types of designs, and a substantial amount of overlap exists in the typologies. For purposes of this discussion I will identify and discuss the six types that my colleagues and I advanced in 2003 (Creswell et al., 2003).

Planning Mixed Methods Procedures

It is helpful, however, before discussing the six types, to consider several aspects that influence the design of procedures for a mixed methods study. Four important aspects are timing, weighting, mixing, and theorizing (as shown in Figure 10.1).

Timing

Proposal developers need to consider the **timing** of their qualitative and quantitative data collection, whether it will be in phases (sequentially) or gathered at the same time (concurrently). When the data are collected in phases, either the qualitative or the quantitative data can come first. It depends on the initial intent of the researcher. When qualitative data are collected first, the intent is to explore the topic with participants at sites. Then the researcher expands the understanding through a second phase in which data are collected from a large number of people (typically a sample representative of a population). When data are collected concurrently, both quantitative and qualitative data are gathered at the same time and the implementation is simultaneous. In many projects it may be unworkable to collect data over an expanded time period (e.g., in the health sciences when busy medical personnel have limited time for data collection in the field). In this case, it is more manageable to collect both quantitative and qualitative data at roughly the same time, when the researcher(s) is in the field collecting data, rather than to revisit the field multiple times for data collection.

Weighting

A second factor that goes into designing procedures is the **weight** or priority given to quantitative or qualitative research in a particular study. In some studies, the weight might be equal; in other studies, it might emphasize

Timing	Weighting	Mixing	Theorizing
No Sequence concurrent	Equal	Integrating	Explicit
Sequential-Qualitative first	Qualitative	Connecting	Implicit
Sequential-Quantitative first	Quantitative	Embedding	

Figure 10.1   Aspects to Consider in Planning a Mixed Methods Design

SOURCE: Adapted from Creswell et al. (2003).

one or the other. A priority for one type depends on the interests of the researcher, the audience for the study (e.g., faculty committee, professional association), and what the investigator seeks to emphasize in the study. In practical terms, weight occurs in a mixed methods study through such strategies as whether quantitative or qualitative information is emphasized first, the extent of treatment of one type of data or the other in the project, or the use of primarily an inductive approach (i.e., generating themes in qualitative) or a deductive approach (i.e., testing a theory). Sometimes the researcher intentionally uses one form of data in a supportive role to a larger study, as is found in some experimental trials (see Rogers et al., 2003).

Mixing

Mixing the data (and in a larger sense, mixing the research questions, philosophy, the interpretation) is difficult at best when one considers that qualitative data consists of text and images and quantitative data, numbers. There are two different questions here: *When* does a researcher mix in a mixed methods study? And *how* does mixing occur? The first question is much easier to answer than the second. Mixing of the two types of data might occur at several stages: the data collection, the data analysis, interpretation, or at all three phases. For proposal developers using mixed methods, it is important to discuss and present in a proposal when the mixing will occur.

How the data are mixed has received considerable recent attention (Creswell & Plano Clark, 2007). **Mixing** means either that the qualitative and quantitative data are actually merged on one end of the continuum, kept separate on the other end of the continuum, or combined in some way between these two extremes. The two data bases might be kept separate but connected; for example, in a two-phase project that begins with a quantitative phase, the analysis of the data and its results can be used to identify participants for qualitative data collection in a follow-up phase. In this situation, the quantitative and qualitative data are connected during the phases of research. **Connected** in mixed methods research means a mixing of the quantitative and qualitative research are connected between a data analysis of the first phase of research and the data collection of the second phase of research. In another study, the researcher might collect both quantitative and qualitative data concurrently and integrate or merge the two databases by transforming the qualitative themes into counts and comparing these counts with descriptive quantitative data. In this case, the mixing consists of **integrating** the two databases by actually merging the quantitative data with the qualitative data. In a final scenario, the researcher might have a primary aim to collect one form of data (say quantitative) and have the other form of data (say qualitative) provide supportive information. Neither integrating the data nor connecting across phases is being utilized. Instead, the researcher is **embedding** a secondary form of data within a larger study having a different form of data as the primary database. The secondary database provides a supporting role in the study.

Theorizing or Transforming Perspectives

A final factor to consider is whether a larger, theoretical perspective guides the entire design. It may be a theory from the social sciences (e.g., adoption theory, leadership theory, attribution theory) or a broad theoretical lens, such as an advocacy/participatory lens (e.g., gender, race, class; see Chapter 3). All researchers bring theories, frameworks and hunches to their inquiries, and these theories may be made explicit in a mixed methods study or be implicit and not mentioned. We will focus here on the use of explicit theories. In mixed methods studies, the theories are found typically in the beginning sections as an orienting lens that shapes the types of questions asked, who participates in the study, how data are collected, and the implications made from the study (typically for change and advocacy). They present an overarching perspective used with all of the mixed methods strategies of inquiry (to be presently discussed). Mertens (2003) provides a good discussion as to how a transforming lens shapes all phases of the research process in mixed methods research.

Alternative Strategies and Visual Models

These four factors—timing, weight, mixing, and theorizing—help to shape the procedures of a mixed methods study. Although these do not exhaust all the possibilities, there are six major strategies for inquirers to choose from in designing a research proposal; they are adapted from Creswell et al. (2003). A proposal would contain a description of the strategy and a visual model of it, as well as basic procedures that the investigator will use in implementing the strategy. Each strategy is briefly described and illustrated in Figures 10.2 and 10.3. The words *qualitative* and *quantitative* have been shortened in the figures to read "qual" and "quan," respectively (see the discussion following the figures).

These mixed methods strategies can be described using notation that has developed in the mixed methods field. **Mixed methods notation** provides shorthand labels and symbols that convey important aspects of mixed methods research, and it provides a way that mixed methods researchers can easily communicate their procedures. The following notation is adapted from Morse (1991), Tashakkori and Teddlie (1998), and Creswell and Plano Clark (2007) who suggest the following:

● A "+" indicates a simultaneous or concurrent form of data collection, with both quantitative and qualitative data collected at same time.

● A "→" indicates a sequential form of data collection, with one form (e.g., qualitative data) building on the other (e.g., quantitative data).

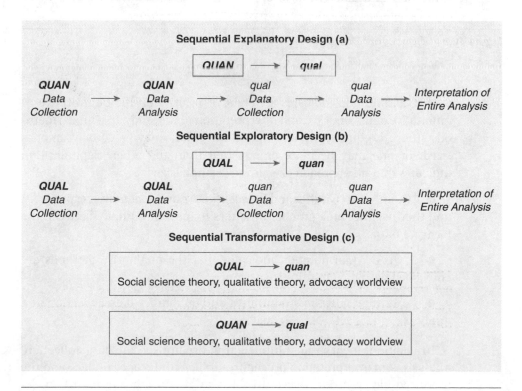

Figure 10.2   Sequential Designs

SOURCE: Adapted from Creswell et al. (2003).

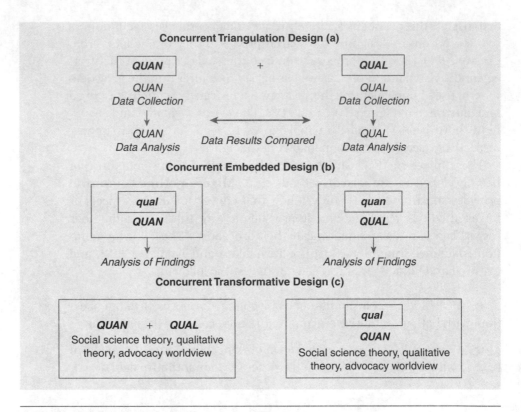

Figure 10.3 Concurrent Designs

SOURCE: Adapted from Creswell et al. (2003).

● Capitalization indicates a weight or priority on the quantitative or qualitative data, analysis, and interpretation in the study. In a mixed methods study, the qualitative and quantitative data may be equally emphasized, or one may be more emphasized than the other. Capitalization indicates that an approach or method is emphasized.

● "Quan" and "Qual" stand for *quantitative* and *qualitative*, respectively, and they use the same number of letters to indicate equality between the forms of data.

● A QUAN/qual notation indicates that the qualitative methods are embedded within a quantitative design.

● Boxes highlight the quantitative and qualitative data collection and analysis.

In addition, incorporated into each figure are specific data collection, analysis, and interpretation procedures to help the reader understand the more specific procedures used. In this way, a figure has at least two elements: the general procedure of mixed methods being used and the more specific procedures of data collection, analysis, and interpretation.

Sequential Explanatory Strategy

The **sequential explanatory strategy** is a popular strategy for mixed methods design that often appeals to researchers with strong quantitative leanings. It is characterized by the collection and analysis of quantitative data in a first phase of research followed by the collection and analysis of qualitative data in a second phase that builds on the results of the initial quantitative results. Weight typically is given to the quantitative data, and the mixing of the data occurs when the initial quantitative results *informs* the secondary qualitative data collection. Thus, the two forms of data are separate but connected. An explicit theory may or may not inform the overall procedure. The steps of this strategy are pictured in Figure 10.2a.

A sequential explanatory design is typically used to explain and interpret quantitative results by collecting and analyzing follow-up qualitative data. It can be especially useful when unexpected results arise from a quantitative study (Morse, 1991). In this case, the qualitative data collection that follows can be used to examine these surprising results in more detail. This strategy may or may not have a specific theoretical perspective. The straightforward nature of this design is one of its main strengths. It is easy to implement because the steps fall into clear, separate stages. In addition, this design feature makes it easy to describe and to report. The main weakness of this design is the length of time involved in data collection, with the two separate phases. This is especially a drawback if the two phases are given equal priority.

Sequential Exploratory Strategy

This next strategy is similar to the explanatory sequential approach except that the phases are reversed. The **sequential exploratory strategy** involves a first phase of qualitative data collection and analysis, followed by a second phase of quantitative data collection and analysis that *builds* on the results of the first qualitative phase. Weight is generally placed on the first phase, and the data are mixed through being connected between the qualitative data analysis and the quantitative data collection. The design may or may not be implemented within an explicit theoretical perspective (see Figure 10.2b).

At the most basic level, the purpose of this strategy is to use quantitative data and results to assist in the interpretation of qualitative findings. Unlike the sequential explanatory approach, which is better suited to explaining and interpreting relationships, the primary focus of this model is to initially explore a phenomenon. Morgan (1998) suggested that this design is appropriate to use when testing elements of an emergent theory resulting from the qualitative phase and that it can also be used to generalize qualitative findings to different samples. Similarly, Morse (1991) cited one purpose for selecting this approach: to determine the distribution of a phenomenon within a chosen population. Finally, the sequential exploratory strategy is often discussed as the procedure of choice when a researcher needs to develop an instrument because existing instruments are inadequate or not available. Using a three-phase approach, the researcher first gathers qualitative data and analyzes it (Phase 1), and uses the analysis to development an instrument (Phase 2) that is subsequently administered to a sample of a population (Phase 3; Creswell & Plano Clark, 2007).

The sequential exploratory strategy has many of the same advantages as the sequential explanatory model. Its two-phase approach (qualitative research followed by quantitative research) makes it easy to implement and straightforward to describe and report. It is useful to a researcher who wants to explore a phenomenon but also wants to expand on the qualitative findings. This model is especially advantageous when a researcher is building a new instrument. In addition, this model could make a largely qualitative study more palatable to an adviser, committee, or research community well versed in quantitative research and that may be unfamiliar with the qualitative approaches. As with the sequential explanatory approach, the sequential exploratory model requires a substantial length of time to complete both data collection phases, which can be a drawback for some research situations. In addition, the researcher has to make some key decisions about which findings from the initial qualitative phase will be focused on in the subsequent quantitative phase (e.g., one theme, comparisons among groups, multiple themes).

Sequential Transformative Strategy

This final sequential approach has two distinct data collection phases, one following the other as in the first two strategies described (see Figure 10.2c). The **sequential transformative strategy** is a two-phase project with a theoretical lens (e.g., gender, race, social science theory) overlaying the sequential procedures. It too has an initial phase (either quantitative or qualitative) followed by a second phase (either qualitative or quantitative) that builds on the earlier phase. The theoretical lens is introduced in the introduction to a proposal, shapes a directional research question aimed at exploring a problem (e.g., inequality, discrimination, injustice), creates sensitivity to collecting data from marginalized or underrepresented groups, and ends with a call for action. In this design the researcher may use either method in the first phase of research, and the weight can be given to either or distributed evenly to both phases. The mixing is connected as in all sequential designs. Unlike the sequential exploratory and explanatory approaches, the sequential transformative model has a theoretical perspective to guide the study. The aim of this theoretical perspective, whether it be a conceptual framework, a specific ideology, or advocacy, is more important in guiding the study than the use of methods alone.

The purpose of a sequential transformative strategy is to best serve the theoretical perspective of the researcher. By using two phases, a sequential transformative researcher may be able to give voice to diverse perspectives, to better advocate for participants, or to better understand a phenomenon or process that is changing as a result of being studied.

The sequential transformative model shares the methodological strengths and weaknesses of the other two sequential approaches. Its use of distinct phases facilitates its implementation, description, and sharing of results, although it also requires the time to complete two data collection phases. More important, this design places mixed methods research within a transformative framework. Therefore, this strategy may be more appealing and acceptable to those researchers already using a transformative framework within one distinct methodology, such as qualitative research. Unfortunately, because little has been written to date on this approach, one weakness is that there is little guidance on how to use the transformative vision to guide the methods. Likewise, as with all sequential strategies, key decisions need to be made about what findings in the first phase will be the focus of the second phase.

Concurrent Triangulation Strategy

The **concurrent triangulation** approach is probably the most familiar of the six major mixed methods models (see Figure 10.3a). In a concurrent triangulation approach, the researcher collects both quantitative and qualitative data concurrently and then compares the two databases to determine if there is convergence, differences, or some combination. Some authors refer to this comparison as *confirmation, disconfirmation, cross-validation,* or *corroboration* (Greene, Caracelli, & Graham, 1989; Morgan, 1998; Steckler, McLeroy, Goodman, Bird, & McCormick, 1992). This model generally uses separate quantitative and qualitative methods as a means to offset the weaknesses inherent within one method with the strengths of the other (or conversely, the strength of one adds to the strength of the other). In this approach, the quantitative and qualitative data collection is concurrent, happening in one phase of the research study. Ideally, the weight is equal between the two methods, but often in practice, priority may be given to one or the other. The mixing during this approach, usually found in an interpretation or discussion section, is to actually merge the data (i.e., transform one type of data to the other type of data so that they can easily be compared) or integrate or compare the results of two databases side by side in a discussion. This side-by-side integration is often seen in published mixed methods studies in which a discussion section first provides quantitative statistical results followed by qualitative quotes that support or disconfirm the quantitative results.

This traditional mixed methods model is advantageous because it is familiar to most researchers and can result in well-validated and substantiated findings. I find that most researchers when they first consider mixed methods employ this model of gathering both quantitative and qualitative data and comparing the two data sources. In addition, the concurrent data collection results in a shorter data collection time period as compared to one of the sequential approaches because both the qualitative and quantitative data are gathered at one time at the research site.

This model also has a number of limitations. It requires great effort and expertise to adequately study a phenomenon with two separate methods. It also can be difficult to compare the results of two analyses using data of different forms. In addition, a researcher may be unclear how to resolve discrepancies that arise in comparing the results, although procedures are emerging in the literature, such as conducting additional data collection to resolve the discrepancy, revisiting the original database, gaining new insight from the disparity of the data, or developing a new project that addresses the discrepancy (Creswell & Plano Clark, 2007).

Concurrent Embedded Strategy

Like the concurrent triangulation approach, the **concurrent embedded** strategy of mixed methods can be identified by its use of one data collection phase, during which both quantitative and qualitative data are collected simultaneously (see Figure 10.3b). Unlike the traditional triangulation model, a concurrent embedded approach has a primary method that guides the project and a secondary database that provides a supporting role in the procedures. Given less priority, the secondary method (quantitative or qualitative) is embedded, or nested, within the predominant method (qualitative or quantitative). This embedding may mean that the secondary method addresses a different question than the primary method (e.g., in an experiment, the quantitative data addresses the outcomes expected from the treatments while the qualitative data explores the processes experienced by individuals in the treatment groups) or seeks information at a different level of analysis (the analogy to hierarchical analysis in quantitative research is helpful in conceptualizing these levels—see Tashakkori and Teddlie, 1998). The mixing of the data from the two methods is often to integrate the information and compare one data source with the other, typically accomplished in a discussion section of a study. However, the data may also not be compared but reside side by side as two different pictures that provide an overall composite assessment of the problem. This would be the case when the researcher uses this approach to assess different research questions or different levels in an organization. Similar to the other approaches, an explicit theoretical perspective can be used in this model, typically to inform the primary method.

The concurrent embedded model may be used to serve a variety of purposes. Often, this model is used so that a researcher can gain broader perspectives as a result of using the different methods as opposed to using the predominant method alone. For example, Morse (1991) noted that a primarily qualitative design could embed some quantitative data to enrich the description of the sample participants. Likewise, she described how qualitative data could be used to describe an aspect of a quantitative study that cannot be quantified. In addition, a concurrent embedded model may be employed when a researcher chooses to utilize different methods to study different groups or levels. For example, if an organization is being studied, employees could be studied quantitatively, managers could be interviewed qualitatively, entire divisions could be analyzed with quantitative data, and so forth. Tashakkori and Teddlie (1998) described this approach as a multilevel design. Finally, one method could be used within a framework of the other method, such as if a researcher designed and conducted an experiment to examine treatment outcomes but used case study methodology to study how participants in the study experienced the treatment procedures.

This mixed methods model is attractive for several reasons. A researcher is able to collect the two types of data simultaneously, during a single data collection phase. It provides a study with the advantages of both quantitative and qualitative data. In addition, by using the two different methods in this fashion, a researcher can gain perspectives from the different types of data or from different levels within the study.

There are also limitations to consider when choosing this approach. The data need to be transformed in some way so that they can be integrated within the analysis phase of the research. In addition, if the two databases are compared, discrepancies may occur that need to be resolved. Because the two methods are unequal in their priority, this approach also results in unequal evidence within a study, which may be a disadvantage when interpreting the final results.

Concurrent Transformative Strategy

As with the sequential transformative model, the **concurrent transformative** approach is guided by the researcher's use of a specific theoretical perspective as well as the concurrent collection of both quantitative and qualitative data (see Figure 10.3c). This perspective can be based on ideologies such as critical theory, advocacy, participatory research, or a conceptual or theoretical framework. This perspective is reflected in the purpose or research questions of the study. It is the driving force behind all methodological choices, such as defining the problem, identifying the design and data sources, analyzing, interpreting, and reporting results. The choice of a concurrent model, whether it is triangulation or embedded design, is made to facilitate this perspective. For example, the design may have one method embedded in the other so that diverse participants are given a voice in the change process of an organization. It may involve a triangulation of quantitative and qualitative data to best converge information to provide evidence for an inequality of policies in an organization.

Thus, the concurrent transformative model may take on the design features of either a triangulation or an embedded approach (the two types of data are collected at the same time during one data collection phase and may have equal or unequal priority). The mixing of the data would be through merging, connecting, or embedding the data. Because the concurrent transformative model shares features with the triangulation and embedded approaches, it also shares their specific strengths and weaknesses. However, this model has the added advantage of positioning mixed methods research within a transformative framework, which may make it especially appealing to those qualitative or quantitative researchers already using a transformative framework to guide their inquiry.

Choosing a Mixed Methods Strategy

Proposal developers need to convey the specific strategy for mixed methods data collection they plan to use. Further, they need to advance a visual figure that presents the procedures they plan to use. Figures 10.2 and 10.3 provide some useful models for guidance. Here are some **research tips** about how to select a mixed methods strategy:

● Use the information in Figure 10.1 to assess the aspects that you will be working with in your mixed methods procedures, and then identify one of the six approaches discussed in this chapter as the primary design for your proposed study. Provide a working definition for this design, along with a visual model and a rationale as to why it is a useful design for you to us.

● Consider the amount of time you have to collect data. Concurrent approaches are less time consuming because both qualitative and quantitative data are collected at the same time in the same visit to the field.

● Remember that the collection and analysis of both quantitative and qualitative data is a rigorous, time-consuming process. When time is a problem, I encourage students to think about an embedded model of design. This model emphasizes a major primary form of data collection (e.g., surveys), and it can include a minor secondary form of data collection (e.g., a few interviews with some of the participants who completed the surveys). The fact that both forms of data are not equal in size and rigor enables the study to be reduced in scope and manageable for the time and resources available.

● Consider using the explanatory sequential approach. It is a favorite of many students, especially those who have little experience with qualitative research and substantial background in quantitative research. In this approach, an initial quantitative data collection is followed by a secondary qualitative data collection to follow up the quantitative results.

● Study published articles that use different approaches and determine which one makes the most sense to you. Creswell and Plano Clark (2007) include four complete journal articles so that readers can examine the detail of studies employing different forms of designs.

● Find a published mixed methods journal article that uses your design and introduce it to your adviser and faculty committee so that they have a working model for the approach you plan to use in your study. Since we are at the early stage of adopting mixed methods research in many fields, a published example of research in your field will help to create both legitimacy for mixed methods research and the idea that it is a feasible approach to research for graduate committees or other audiences.

DATA COLLECTION PROCEDURES

Although the visual model and the discussion about the specific strategies in a proposal provide a picture of the procedures, it is helpful to discuss the specific types of data to be collected. It is also important to identify the sampling strategies and the approaches used to establish validity of the data.

● Identify and be specific about the type of data—both quantitative and qualitative—that will be collected during the proposed study. Refer to Table 1.3, which shows both quantitative and qualitative data. Data differs in terms of open-ended versus closed-ended responses. Some forms of data, such as interviews and observations, can be either quantitative or qualitative depending on how open (qualitative) or closed (quantitative) the response options might be in an interview or a checklist for an observation. Although reducing information to numbers is the approach used in quantitative research, it is also used in qualitative research.

● Recognize that quantitative data often involve random sampling, so that each individual has an equal probability of being selected, and the sample can be generalized to the larger population. In qualitative data collection, purposeful sampling is used so that individuals are selected because they have experienced the central phenomenon. Rigorous sampling procedures need to be conveyed in a proposal for the quantitative and qualitative data collection. In addition, Teddlie and Yu (2007) have developed a typology of five types of mixed methods sampling which relates sampling to the strategies for mixed methods that I have discussed:

  • Basic strategies that involve combining quantitative and qualitative sampling (e.g., stratified purposeful sampling, purposive random sampling)

  • Sequential sampling, in which the sampling from the first phase or strand informs the second phase or strand

- Concurrent sampling, in which quantitative probability and qualitative purposeful sampling are combined as independent sampling procedures or jointly (e.g., a survey with both closed-ended and open-ended responses)

- Multilevel sampling, in which sampling occurs in two or more levels or units of analysis

- Sampling using any combination of the foregoing strategies

● Include detailed procedures in your visual model. For example, in a sequential explanatory model, the general procedures are higher on the page and the detailed procedures below them, as shown in Figure 10.2a. However, the figure can be detailed even further. For example, a discussion of this approach might include describing the use of survey data collection followed by both descriptive and inferential data analysis in the first phase. Then qualitative observations and coding and thematic analysis within an ethnographic design might be mentioned for the second phase.

DATA ANALYSIS AND VALIDATION PROCEDURES

Data analysis in mixed methods research relates to the type of research strategy chosen for the procedures. Thus, in a proposal, the procedures need to be identified within the design. However, analysis occurs both within the quantitative (descriptive and inferential numeric analysis) and the qualitative (description and thematic text or image analysis) approach and often between the two approaches. For example, some of the more popular mixed methods data analysis approaches are the following (see Caracelli & Greene, 1993; Creswell & Plano Clark, 2007; Tashakkori & Teddlie, 1998):

● Data transformation: In the concurrent strategies, a researcher may quantify the qualitative data. This involves creating codes and themes qualitatively, then counting the number of times they occur in the text data (or possibly the extent of talk about a code or theme by counting lines or sentences). This quantification of qualitative data then enables a researcher to compare quantitative results with the qualitative data. Alternatively, an inquirer may qualify quantitative data. For instance, in a factor analysis of data from a scale on an instrument, the researcher may create factors or themes that then can be compared with themes from the qualitative database.

● Explore outliers: In a sequential model, an analysis of quantitative data in the first phase can yield extreme or outlier cases. Follow-up qualitative interviews with these outlier cases can provide insight about why they diverged from the quantitative sample.

● Instrument development: In a sequential approach, obtain themes and specific statements from participants in an initial qualitative data collection. In the next phase, use these statements as specific items and the themes for scales to create a survey instrument that is grounded in the views of the participants. A third, final phase might be to validate the instrument with a large sample representative of a population.

● Examine multiple levels: In a concurrent embedded model, conduct a survey at one level (e.g., with families) to gather quantitative results about a sample. At the same time, collect qualitative interviews (e.g., with individuals) to explore the phenomenon with specific individuals in the families.

● Create a matrix: When comparing data in a concurrent type of approach, combine information from both the quantitative and qualitative data collection into a matrix. The horizontal axis of this matrix could be a quantitative categorical variable (e.g., type of provider—nurse, physician, and medical assistant) and the vertical axis would be the qualitative data (e.g., five themes about caring relationships between providers and patients). Information in the cells could be either quotes from the qualitative data, counts of the number of codes from the qualitative data, or some combination. In this way, the matrix would present an analysis of the combined qualitative and quantitative data. Qualitative computer software programs provide matrix output capabilities for the mixed methods researcher (see Chapter 9).

Another aspect of data analysis in mixed methods research to describe in a proposal is the series of steps taken to check the validity of both the quantitative data and the accuracy of the qualitative findings. Writers on mixed methods advocate for the use of validity procedures for both the quantitative and qualitative phases of the study (Tashakkori & Teddlie, 1998). The proposal writer discusses the validity and reliability of the scores from past uses of instruments employed in the study. In addition, potential threats to internal validity for experiments and surveys are noted (see Chapter 8). For the qualitative data, the strategies that will be used to check the accuracy of the findings need to be mentioned (see Chapter 9). These may include triangulating data sources, member checking, detailed description, or other approaches.

An emerging field of study is to consider how validity might be different for mixed methods studies than for a quantitative or a qualitative study. Writers have begun to develop a bilingual nomenclature for mixed methods research and have called validity *legitimation* (Onwuegbuzie & Johnson, 2006, p. 55). The legitimation of the mixed methods study relates to many phases of the research process, from philosophical issues (e.g., are the philosophical positions blended into a usable form?) to inferences drawn (e.g., yield high-quality inferences) and to the value of the study for consumers (see Onwuegbuzie & Johnson, 2006). For individuals writing a mixed methods research proposal, consider the types of validity associated with the quantitative component (see Chapter 8), validity related to the qualitative strand (see Chapter 9), and any validity issues that might arise that relate to the mixed methods approach. Validity issues in mixed methods research may relate to the types of strategies discussed in this chapter. These may relate to sample selection, sample size, follow up on contradictory results, bias in data collection, inadequate procedures, or the use of conflicting research questions (see Creswell & Plano Clark, 2007 for a discussion of these ideas).

REPORT PRESENTATION STRUCTURE

The structure for the report, like the data analysis, follows the type of strategy chosen for the proposed study. Because mixed methods studies may not be familiar to audiences, it is helpful to provide some guidance as to how to structure the final report.

● For a sequential study, mixed methods researchers typically organize the report of procedures into quantitative data collection and quantitative data analysis followed by qualitative data and collection and analysis. Then, in the conclusions or interpretation phase of the study, the researcher comments on how the qualitative findings helped to elaborate on or extend the quantitative results. Alternatively, the qualitative data collection and analysis could come first, followed by the quantitative data collection and analysis. In either structure, the writer typically presents the project as two distinct phases, with separate headings for each phase.

● In a concurrent study, the quantitative and qualitative data collection may be presented in separate sections, but the analysis and interpretation combines the two forms of data to seek convergence or similarities among the results. The structure of this type of mixed methods study does not make a clear distinction between the quantitative and qualitative phases.

● In a transformative study, the structure typically involves advancing the advocacy issue in the beginning and then using either the sequential or concurrent structure as a means of organizing the content. In the end, a separate section may advance an agenda for change or reform that has developed as a result of the research.

EXAMPLES OF MIXED METHODS PROCEDURES

Illustrations follow of mixed methods studies that use both the sequential and concurrent strategies and procedures.

Example 10.1   *A Sequential Strategy of Inquiry*

Kushman (1992) studied two types of teacher workplace commitment—organizational commitment and commitment to student learning—in 63 urban elementary and middle schools. He posed a two-phase explanatory sequential mixed methods study, as presented in the purpose statement:

> The central premise of this study was that organizational commitment and commitment to student learning address distinct but equally important teacher attitudes for an organizationally effective school, an idea that has some support in the literature but requires further empirical validation. . . . Phase 1 was a quantitative study that looked at statistical relationships between teacher commitment and organizational antecedents and outcomes in elementary and middle schools. Following this macrolevel analysis, Phase 2 looked within specific schools, using qualitative/case study methods to better understand the dynamics of teacher commitment.
>
> (Kushman, 1992, p. 13)

This purpose statement illustrates the combination of a purpose with the rationale for mixing ("to better understand") as well as the specific types of data collected during the study. The introduction focused on the need to examine organizational commitment and commitment to student learning, leading to a priority for the quantitative approach. This priority was further illustrated in sections defining organizational commitment and commitment to student learning and the use of extensive literature to document these two concepts. A conceptual framework then followed, complete with a visual model, and research questions were posed to explore relationships. This provided a theoretical lead for the quantitative phase of the study (Morse, 1991). The implementation was QUAN→qual in this two-phase study. The author presented results in two phases, with the first—the quantitative results—displaying and discussing correlations, regressions, and two-way ANOVAs. Then the case study results were presented in terms of themes and subthemes supported by quotations. The mixing of the quantitative results and qualitative findings occurred in the final discussion, in which the researcher highlighted the quantitative results and the complexities that surfaced from the qualitative results. The author did not use a theoretical perspective as a lens in the study.

Example 10.2   *A Concurrent Strategy of Inquiry*

In 1993, Hossler and Vesper conducted a study examining the factors associated with parental savings for children attending higher education campuses.

(Continued)

(Continued)

Using longitudinal data collected from students and parents over a 3-year period, the authors examined factors most strongly associated with parental savings for postsecondary education. Their results found that parental support, educational expectations, and knowledge of college costs were important factors. Most important, for our purposes, the authors collected information from parents and students on 182 surveys and from 56 interviews. Their purpose indicated an interest in triangulating the findings:

> In an effort to shed light on parental saving, this article examines parental saving behaviors. Using student and parent data from a longitudinal study employing multiple surveys over a three-year period, logistic regression was used to identify the factors most strongly associated with parental saving for postsecondary education. In addition, insights gained from the interviews of a small sub-sample of students and parents who were interviewed five times during the three-year period are used to further examine parental savings.
>
> (Hossler & Vesper, 1993, p. 141)

The actual data was collected from 182 student and parent participants from surveys over a 4-year period of time and from 56 students and their parents in interviews. From the purpose statement, we can see that they collected data concurrently as an implementation strategy. They provide extensive discussion of the quantitative analysis of the survey data, including a discussion about the measurement of variables and the details of the logistic regression data analysis. They also mention the limitations of the quantitative analysis and specific t-test and regression results. In contrast, they devote one page to the qualitative data analysis and note briefly the themes that occurred in the discussion. The weight in this mixed methods study was assigned to quantitative data collection and analysis, and the notation for the study would be QUAN + qual. The mixing of the two data sources occurred in a section titled "Discussion of Survey and Interview Results" (p. 155), at the interpretation stage of the research process. In this section, they compared the importance of the factors explaining parental savings for the quantitative results, on one hand, with the findings from the interview data on the other. Similar to Example 10.1, no theoretical lens guided the study, although the article began with the literature on econometric studies and research on college choice and ended with an "Augmented Model of Parental Savings." Thus, we might characterize the use of theory in this study as inductive (as in qualitative inquiry), drawn from the literature (as in quantitative research), and ultimately as generated during the process of research.

Example 10.3   *A Transformative Strategy of Inquiry*

A feminist lens was used in a transformative triangulation mixed methods study by Bhopal (2000). She was interested in examining whether theories of patriarchy apply to South Asian women (from India, Pakistan and Bangladesh) living in East London. Since these women often have arranged marriages and are given dowries, she assumed that forms of patriarchy exist for them that do not exist for White women in Britain. Her overall aim was to "build up a detailed knowledge of women's lives, their feelings towards their own roles and what they actually did in the home, their attitudes towards arranged marriages, dowries, domestic labour and domestic finance" (p. 70). She studied 60 women using qualitative and quantitative methods,

> (to) investigate the varying significance of difference through which women experienced patriarchy . . . and to present accurate information regarding the number of women who experienced different forms of patriarchy and test for the strength of associations between different influences of patriarchy.
>
> (Bhopal, 2000, p. 68)

She found that education had a significant impact upon the women's lives. Moreover, her discussion presented how feminist methodologies informed her study. She discussed how she addressed women's lives in their own terms, using the language and categories in which women express themselves. Her research did not just involve women, it was *for* women. Her research also involved putting herself into the process of production in which she made explicit her reasoning procedures and was self-reflexive about her own perceptions and biases. In this way, her mixed methods study helped to expose the lives of women and have a transformative effect in which the "women benefited from the research project" (p. 76).

The intent of the author was to provide a voice for women and to give a more powerful voice to gender inequality. The quantitative data provided the generalized patterns of participation while the qualitative data provided the personal narratives of women. The timing of the study was to first collect survey data and then interview women to follow up and understand their participation in more depth (an explanatory sequential design). The weight of the qualitative and quantitative components was equal, with the thought that both contribute to understanding the research problem. The mixing was through connecting the results from the quantitative survey and exploring these in more depth in the qualitative phase. Because feminist theory was discussed throughout the article with a focus on equality and giving voice to women, the study employed an explicit theoretical feminist lens.

SUMMARY

In designing the procedures for a mixed methods study, begin by conveying the nature of mixed methods research. This includes tracing its history, defining it, and mentioning its applications in many fields of research. Then, state and employ four criteria to select an appropriate mixed methods strategy. Indicate the timing strategy for data collection (concurrent or sequential). Also state weight or priority given to the quantitative or qualitative approach, such as equal weight, or a priority to quantitative or qualitative data. Mention how the data will be mixed, such as through merging the data, connecting the data from one phase to another, or embedding a secondary source of data into a larger, primary source. Finally, identify whether a theoretical lens or framework will guide the study, such as a theory from the social sciences or a lens from an advocacy perspective (e.g., feminism, racial perspective).

Six strategies are organized around whether the data are collected sequentially (explanatory and exploratory), concurrently (triangulation and nested), or with a transformative lens (sequential or concurrent). Each model has strengths and weaknesses, although the sequential approach is the easiest to implement. Choice of strategy also can be presented in a figure in the research proposal. Then, specific procedures can be related to the figure to help the reader understand the flow of activities in a project. These include the types of quantitative and qualitative data to be collected as well as the procedures for data analysis. Typically, data analysis involves data transformation, exploring outliers, examining multiple levels, or creating matrices that combine the quantitative results and the qualitative findings. Validity procedures also need to be explicitly described. The final written report, because it may be unfamiliar to audiences, can also be described in a proposal. Each of the three types of strategies—sequential, concurrent, and transformative—has a different structural approach to writing a mixed methods study.

Writing Exercises

1. Design a combined qualitative and quantitative study that employs two phases sequentially. Discuss and provide a rationale for why the phases are ordered in the sequence you propose.

2. Design a combined qualitative and quantitative study that gives priority to qualitative data collection and less priority to quantitative data collection. Discuss the approach to be taken in writing the introduction, the purpose statement, the research questions, and the specific forms of data collection.

3. Develop a visual figure and specific procedures that illustrate the use of a theoretical lens, such as a feminist perspective. Use the procedures of either a sequential or concurrent model for conducting the study. Use appropriate notation in the figure.

ADDITIONAL READINGS

Creswell, J. W., & Plano Clark, V. L. (2007). *Designing and conducting mixed methods research.* Thousand Oaks, CA: Sage.

Plano Clark, V. L. & Creswell, J. W. (2008). *The mixed methods reader.* Thousand Oaks, CA: Sage.

Creswell and Plano Clark have developed two books that provide an introduction to mixed methods research and sample research studies and methodological articles about mixed methods research. In the first, we emphasize four types of mixed methods designs: explanatory sequential, exploratory sequential, triangulation, and the embedded designs and provide articles that illustrate each design. This design theme is carried forward in their *Reader*, in which additional articles are included that present actual research studies employing the designs as well as discussions surrounding the basic ideas of the four designs.

Greene, J. C., Caracelli, V. J., & Graham, W. F. (1989). Toward a conceptual framework for mixed-method evaluation designs. *Educational Evaluation and Policy Analysis, 11*(3), 255–274.

Jennifer Greene and associates undertook a study of 57 mixed methods evaluation studies reported from 1980 to 1988. From this analysis, they developed five different mixed methods purposes and seven design characteristics. They found the purposes of mixed methods studies to be based on seeking convergence (triangulation), examining different facets of a phenomenon (complementarity), using the methods sequentially (development), discovering paradox and fresh perspectives (initiation), and adding breadth and scope to a project (expansion). They also found that the studies varied in terms of the assumptions, strengths, and limitations of the method and whether they addressed different phenomena or the same phenomena; were implemented within the same or different paradigms; were given equal or different weight in the study; and were implemented independently, concurrently, or sequentially. Using the purposes and the design characteristics, the authors recommended several mixed methods designs.

Morse, J. M. (1991). Approaches to qualitative-quantitative methodological triangulation. *Nursing Research, 40*(1), 120–123.

Janice Morse suggests that using qualitative and quantitative methods to address the same research problem leads to issues of weighing each method and their sequence in a study. Based on these ideas, she then advances two forms of methodological triangulation: simultaneous, using both methods at the same time; and sequential, using the results of one method for planning the next method. These two forms are described using a notation of capital and lowercase letters that signify relative weight as well as sequence. The different approaches to triangulation are then discussed in the light of their purpose, limitations, and approaches.

Tashakkori, A., & Teddlie, C. (Eds.). (2003). *Handbook of mixed methods in the social & behavioral sciences.* Thousand Oaks, CA: Sage.

This handbook, edited by Abbas Tashakkori and Charles Teddlie, represents the most substantial effort to date to bring together the leading writers about mixed methods research. In 27 chapters, it introduces the reader to mixed methods, illustrates methodological and analytic issues in its use, identifies applications in the social and human sciences, and plots future directions. For example, separate chapters illustrate the use of mixed methods research in evaluation, management and organization, health sciences, nursing, psychology, sociology, and education.

Glossary

Abstract in a literature review is a brief review of the literature (typically in a short paragraph) that summarizes major elements to enable a reader to understand the basic features of the article.

Advocacy/participatory worldview is a philosophy of research in which inquiry is intertwined with politics and a political agenda. Thus, the research contains an action agenda for reform that may change the lives of the participants, the institutions in which individuals work or live, and the researcher's life. Moreover, specific issues are addressed that speak to important social issues of the day, such as empowerment, inequality, oppression, domination, suppression, and alienation.

Attention or interest thoughts in writing are sentences whose purposes are to keep the reader on track, organize ideas, and keep an individual's attention.

Big thoughts in writing are sentences containing specific ideas or images that fall within the realm of umbrella thoughts and serve to reinforce, clarify, or elaborate upon the umbrella thoughts.

Case studies are a qualitative strategy in which the researcher explores in depth a program, event, activity, process, or one or more individuals. The case(s) are bounded by time and activity, and researchers collect detailed information using a variety of data collection procedures over a sustained period of time.

Central question in qualitative research is a broad question posed by the researcher that asks for an exploration of the central phenomenon or concept in a study.

Codes of ethics are the ethical rules and principles drafted by professional associations that govern scholarly research in the disciplines.

Coding is the process of organizing the material into chunks or segments of text in order to develop a general meaning of each segment.

Coherence in writing means that the ideas tie together and logically flow from one sentence to another and from one paragraph to another.

Computer databases of the literature are now available in libraries, and they provide quick access to thousands of journals, conference papers, and materials.

Concurrent embedded strategy of mixed methods research can be identified by its use of one data collection phase, during which both quantitative and qualitative data are collected simultaneously. Unlike the traditional triangulation model, a concurrent embedded approach has a primary method that guides the project and a secondary method that provides a supporting role in the procedures.

Concurrent mixed methods procedures are those in which the researcher converges or merges quantitative and qualitative data in order to provide a comprehensive analysis of the research problem.

Concurrent transformative approach in mixed methods is guided by the researcher's use of a specific theoretical perspective as well as the concurrent collection of both quantitative and qualitative data.

Concurrent triangulation strategy in mixed methods is an approach in which the researcher collects both quantitative and qualitative data concurrently and then compares the two databases to determine if there is convergence, differences, or some combination.

Confidence interval is an estimate in quantitative research of the range of upper and lower statistical values that are consistent with the observed data and are likely to contain the actual population mean.

Connected in mixed methods research means that the quantitative and qualitative research are connected between a data analysis of the first phase of research and data collection of the second phase.

Construct validity occurs when investigators use adequate definitions and measures of variables.

Deficiencies in past literature may exist because topics have not been explored with a particular group, sample, or population; the literature may need to be replicated or repeated to see if the same findings hold, given new samples of people or new sites for study; or the voice of underrepresented groups have not been heard in published literature.

Definition of terms is a section that may be found in a research proposal that defines terms that readers may not understand.

Descriptive analysis of data for variables in a study includes describing the results through means, standard deviations, and range of scores.

Directional hypothesis, as used in quantitative research, is one in which the researcher makes a prediction about the expected direction or outcomes of the study.

Effect size identifies the strength of the conclusions about group differences or the relationships among variables in quantitative studies.

Embedding is a form of mixing in mixed methods research, where a secondary form of data is lodged within a larger study with a different form of data as the primary database. The secondary database provides a supporting role.

Ethnography is a qualitative strategy in which the researcher studies an intact cultural group in a natural setting over a prolonged period of time by collecting primarily observational and interview data.

Experimental design in quantitative research tests the impact of a treatment (or an intervention) on an outcome, controlling for all other factors that might influence that outcome.

Experimental research seeks to determine if a specific treatment influences an outcome in a study. This impact is assessed by providing a specific treatment to one group and withholding it from another group and then determining how both groups score on an outcome.

External validity threats arise when experimenters draw incorrect inferences from the sample data to other persons, other settings, and past or future situations.

Fat in writing refers to words added to prose that are unnecessary to convey the intended meaning.

Gatekeepers are individuals at research sites that provide access to the site and allow or permit a qualitative research study to be undertaken.

Grounded theory is a qualitative strategy in which the researcher derives a general, abstract theory of a process, action, or interaction grounded in the views of participants in a study.

Habit of writing is writing in a regular and continuous way on a proposal rather than in binges or in on-and-off again times.

Inferential questions or hypotheses relate variables or compare groups in terms of variables so that inferences can be drawn from the sample to a population.

Informed consent forms are those that participants sign before they engage in research. This form acknowledges that participants' rights will be protected during data collection.

Integrating the two databases in mixed methods research means that the quantitative and qualitative databases are actually merged through a comparison approach or through data transformation.

Intercoder agreement (or cross-checking) is when two or more coders agree on codes used for the same passages in the text. (It is not that they code the same text, but whether another coder would code a similar passage with the same or a similar code). Statistical procedures or reliability subprograms in qualitative computer software packages can be used to determine the level of consistency of coding.

Internal validity threats are experimental procedures, treatments, or experiences of the participants that threaten the researcher's ability to draw correct inferences from the data about the population in an experiment.

Interpretation of the results in quantitative research means that the researcher draws conclusions from the results for the research questions, hypotheses, and the larger meaning of the results.

Interpretation in qualitative research means that the researcher draws meaning from the findings of data analysis. This meaning may result in lessons learned, information to compare with the literature, or personal experiences.

Interview protocol is a form used by a qualitative researcher for recording and writing down information obtained during an interview.

Literature map is a visual picture (or figure) of the research literature on a topic that illustrates how a particular study contributes to the literature.

Matching participants in experimental research is a procedure in which participants with certain traits or characteristics are matched and then randomly assigned to control and experimental groups.

Mixed methods notation provides shorthand labels and symbols that convey important aspects of mixed methods research, and it provides a way that mixed methods researchers can easily communicate their procedures.

Mixed methods purpose statements contain the overall intent of the study, information about both the quantitative and qualitative strands of the study, and a rationale of incorporating both strands to study the research problem.

Mixed methods research is an approach to inquiry that combines or associates both qualitative and quantitative forms of research. It involves philosophical assumptions, the use of qualitative and quantitative approaches, and the mixing of both approaches in a study.

Mixed methods research question is a special question posed in a mixed methods study that directly addresses the mixing of the quantitative and qualitative strands of the research. This is the question that will be answered in the study based on the mixing.

Mixing means either that the qualitative and quantitative data are actually merged, one end of the continuum, or kept separate, the other end of the continuum, or combined in some way on the continuum.

Narrative hook is a term drawn from English composition, meaning words that are used in the opening sentence of an introduction that serve to draw, engage, or hook the reader into the study.

Narrative research is a qualitative strategy in which the researcher studies the lives of individuals and asks one or more individuals to provide stories about their lives. This information is then often retold or restoried by the researcher into a narrative chronology.

Nondirectional hypothesis in a quantitative study is one in which the researcher makes a prediction, but the exact form of differences (e.g., higher, lower, more, or less) is not specified because the researcher does not know what can be predicted from past literature.

Null hypothesis in quantitative research represents the traditional approach to writing hypotheses: It makes a prediction that, in the general population, no relationship or no significant difference exists between groups on a variable.

Observational protocol is a form used by a qualitative researcher for recording and writing down information while observing.

Phenomenological research is a qualitative strategy in which the researcher identifies the essence of human experiences about a phenomenon as described by participants in a study.

Postpositivists reflect a deterministic philosophy about research in which causes probably determine effects or outcomes. Thus, the problems studied by postpositivists reflect issues that need to identify and assess the causes that influence the outcomes, such as found in experiments.

Pragmatism as a worldview or philosophy arises out of actions, situations, and consequences rather than antecedent conditions (as in postpositivism). There is a concern with applications—what works—and solutions to problems. Instead of focusing on methods, researchers emphasize the research problem and use all approaches available to understand the problem.

Purpose statement in a research proposal sets the objectives, the intent, and the major idea for the study.

Purposefully select participants or sites (or documents or visual material) means that qualitative researchers select individuals who will best help them understand the research problem and the research questions.

Qualitative audio and visual materials take the forms of photographs, art objects, videotapes, or any forms of sound.

Qualitative codebook is a means for organizing qualitative data using a list of predetermined codes that are used for coding the data. This codebook might be composed with the names of codes in one column, a definition of codes in another column, and then specific instances (e.g., line numbers) in which the code is found in the transcripts.

Qualitative documents are public documents (e.g., newspapers, minutes of meetings, official reports) or private documents (e.g., personal journals and diaries, letters, e-mails).

Qualitative generalization is a term that is used in a limited way in qualitative research, since the intent of this form of inquiry is not to generalize findings to individuals, sites, or places outside of those under study. Generalizing findings to theories is an approach used in multiple case study qualitative research, but the researcher needs to have well-documented procedures and a well-developed qualitative database.

Qualitative interviews means that the researcher conducts face-to-face interviews with participants, interviews participants by telephone, or engages in focus group interviews with six to eight interviewees in each group. These interviews involve unstructured and generally open-ended questions that are few in number and intended to elicit views and opinions from the participants.

Qualitative observation means that the researcher takes field notes on the behavior and activities of individuals at the research site and records observations.

Qualitative purpose statements contain information about the central phenomenon explored in the study, the participants in the study, and the research site. It also conveys an emerging design and research words drawn from the language of qualitative inquiry.

Qualitative reliability indicates that a particular approach is consistent across different researchers and different projects.

Qualitative research is a means for exploring and understanding the meaning individuals or groups ascribe to a social or human problem. The process of research involves emerging questions and procedures; collecting data in the participants' setting; analyzing the data inductively, building from particulars to general themes; and making interpretations of the meaning of the data. The final written report has a flexible writing structure.

Qualitative validity means that the researcher checks for the accuracy of the findings by employing certain procedures.

Quantitative hypotheses are predictions the researcher makes about the expected relationships among variables.

Quantitative purpose statements include the variables in the study and their relationship, the participants in a study, and the site for the research. It also includes language associated with quantitative research and the deductive testing of relationships or theories.

Quantitative research is a means for testing objective theories by examining the relationship among variables. These variables can be measured, typically on instruments, so that numbered data can be analyzed using statistical procedures. The final written report has a set structure consisting of introduction, literature and theory, methods, results, and discussion.

Quantitative research questions are interrogative statements that raise questions about the relationships among variables that the investigator seeks to answer.

Quasi-experiment is a form of experimental research in which individuals are not randomly assigned to groups.

Random sampling is a procedure in quantitative research for selecting participants. It means that each individual has an equal probability of being selected from the population, ensuring that the sample will be representative of the population.

Reflexivity means that researchers reflect about how their biases, values, and personal background, such as gender, history, culture, and socioeconomic status, shape their interpretations formed during a study.

Reliability refers to whether scores to items on an instrument are internally consistent (i.e., Are the item responses consistent across constructs?), stable over time (test–retest correlations), and whether there was consistency in test administration and scoring.

Research designs are plans and the procedures for research that span the decisions from broad assumptions to detailed methods of data collection and analysis. It involves the intersection of philosophical assumptions, strategies of inquiry, and specific methods.

Research methods involve the forms of data collection, analysis, and interpretation that researchers propose for their studies.

Research problems are the problems or issues that lead to the need for a study.

Research tips are my thoughts about approaches or techniques that have worked well for me as an experienced researcher.

Response bias is the effect of nonresponses on survey estimates, and it means that if nonrespondents had responded, their responses would have substantially changed the overall results of the survey.

Reviewing studies in an introduction justifies the importance of the study and creates distinctions between past studies and a proposed study.

Script, as used in this book, is a template of a few sentences that contains the major words and ideas for particular parts of a research proposal or report (e.g., purpose statement or research question) and provides space for researchers to insert information that relates to their projects.

Sequential explanatory strategy in mixed methods research is characterized by the collection and analysis of quantitative data in a first phase followed by the collection and analysis of qualitative data in a second phase that builds on the results of the initial quantitative results.

Sequential exploratory strategy in mixed methods research involves a first phase of qualitative data collection and analysis followed by a second phase of quantitative data collection and analysis that builds on the results of the first quantitative phase.

Sequential mixed methods procedures are those in which the researcher seeks to elaborate on or expand on the findings of one method with another method.

Sequential transformative strategy in mixed methods research is a two-phase project with a theoretical lens (e.g., gender, race, social science theory) overlaying the procedures, with an initial phase (either quantitative or qualitative) followed by a second phase (either qualitative or quantitative) that builds on the earlier phase.

Significance of the study in an introduction conveys the importance of the problem for different audiences that may profit from reading and using the study.

Single-subject design or N of 1 design involves observing the behavior of a single individual (or a small number of individuals) over time.

Social constructivists hold the assumption that individuals seek understanding of the world in which they live and work. Individuals develop subjective meanings of their experiences, meanings directed toward certain objects or things.

Statistical conclusion validity arises when experimenters draw inaccurate inferences from the data because of inadequate statistical power or the violation of statistical assumptions.

Strategies of inquiry are types of qualitative, quantitative, and mixed methods designs or models that provide specific direction for procedures in a research design.

Style manuals provide guidelines for creating a scholarly style of a manuscript, such as a consistent format for citing references, creating headings, presenting tables and figures, and using nondiscriminatory language.

Survey design provides a plan for a quantitative or numeric description of trends, attitudes, or opinions of a population by studying a sample of that population.

Survey research provides a quantitative or numeric description of trends, attitudes, or opinions of a population by studying a sample of that population.

Theoretical lens or perspective in qualitative research provides an overall orienting lens that is used to study questions of gender, class, and race (or other issues of marginalized groups). This lens becomes an advocacy perspective that shapes the types of questions asked, informs how data are collected and analyzed, and provides a call for action or change.

Theories in mixed methods research provide an orienting lens that shapes the types of questions asked, who participates in the study, how data are collected, and the implications made from the study (typically for change and advocacy). They present an overarching perspective used with other strategies of inquiry.

Theory in quantitative research is the use of an interrelated set of constructs (or variables) formed into propositions, or hypotheses, that specify the relationship among variables (typically in terms of magnitude or direction) and predicts the outcomes of a study.

Theory use in mixed methods studies may include theory deductively in quantitative theory testing and verification or inductively, as in an emerging qualitative theory or pattern.

Timing in mixed methods research involves collecting the quantitative and qualitative data in phases (sequentially) or gathering it at the same time (concurrently).

Topic is the subject or subject matter of a proposed study that a researcher identifies early in the preparation of a study.

Transformative mixed methods procedures are those in which the researcher uses a theoretical lens (see Chapter 3) as an overarching perspective within a design that contains both quantitative and qualitative data.

True experiment is a form of experimental research in which individuals are randomly assigned to groups.

Validity in quantitative research refers to whether one can draw meaningful and useful inferences from scores on particular instruments.

Validity strategies in qualitative research are procedures (e.g., member checking, triangulating data sources) that qualitative researchers use to demonstrate the accuracy of their findings and convince readers of this accuracy.

Variable refers to a characteristic or attribute of an individual or an organization that can be measured or observed and that varies among the people or organization being studied. A variable typically will vary in two or more categories or on a continuum of scores, and it can be measured.

Weight in mixed methods research is the priority given to quantitative or qualitative research in a particular study. In some studies, the weight might be equal; in others, it might emphasize qualitative or quantitative data.

Worldview is defined as "a basic set of beliefs that guide action" (Guba, 1990, p. 17).

References

Aikin, M. C. (Ed.). (1992). *Encyclopedia of educational research* (6th ed.). New York: Macmillan.

American Psychological Association. (2001). *Publication Manual of the American Psychological Association* (5th ed.). Washington, DC: Author.

Anderson, E. H., & Spencer, M. H. (2002). Cognitive representation of AIDS. *Qualitative Health Research, 12*(10), 1338–1352.

Annual review of psychology. (1950–). Stanford, CA: Annual Reviews.

Ansorge, C., Creswell, J. W., Swidler, S., & Gutmann, M. (2001). *Use of iBook laptop computers in teacher education.* Unpublished manuscript, University of Nebraska-Lincoln.

Asmussen, K. J., & Creswell, J. W. (1995). Campus response to a student gunman. *Journal of Higher Education, 66,* 575–591.

Babbie, E. (1990). *Survey research methods* (2nd ed.). Belmont, CA: Wadsworth.

Babbie, E. (2007). *The practice of social research* (11th ed.). Belmont, CA: Wadsworth/ Thomson.

Bailey, E. P. (1984). *Writing clearly: A contemporary approach.* Columbus, OH: Charles Merrill.

Bausell, R. B. (1994). *Conducting meaningful experiments.* Thousand Oaks, CA: Sage.

Bean, J., & Creswell, J. W. (1980). Student attrition among women at a liberal arts college. *Journal of College Student Personnel, 3,* 320–327.

Beisel, N. (February, 1990). Class, culture, and campaigns against vice in three American cities, 1872–1892. *American Sociological Review, 55,* 44–62.

Bem, D. (1987). Writing the empirical journal article. In M. Zanna & J. Darley (Eds.), *The compleat academic: A practical guide for the beginning social scientist* (pp. 171–201). New York: Random House.

Berg, B. L. (2001). *Qualitative research methods for the social sciences* (4th ed.). Boston: Allyn & Bacon.

Berger, P. L., & Luekmann, T. (1967). *The social construction of reality: A treatise in the sociology of knowledge.* Garden City, NJ: Anchor.

Bhopal, K. (2000). Gender, "race" and power in the research process: South Asian women in East London. In C. Truman, D. M. Mertens, & B. Humphries (Eds.), *Research and inequality.* London: UCL Press.

Blalock, H. (1969). *Theory construction: From verbal to mathematical formulations.* Englewood Cliffs, NJ: Prentice Hall.

Blalock, H. (1985). *Casual models in the social sciences.* New York: Aldine.

Blalock, H. (1991). Are there any constructive alternatives to causal modeling? *Sociological Methodology, 21,* 325–335.

Blase, J. J. (1989, November). The micropolitics of the school: The everyday political orientation of teachers toward open school principals. *Educational Administration Quarterly, 25*(4), 379–409.

Boeker, W. (1992). Power and managerial dismissal: Scapegoating at the top. *Administrative Science Quarterly, 37,* 400–421.

Bogdan, R. C., & Biklen, S. K. (1992). *Qualitative research for education: An introduction to theory and methods.* Boston: Allyn & Bacon.

Boice, R. (1990). *Professors as writers: A self-help guide to productive writing.* Stillwater, OK: New Forums.

Boneva, B., Kraut, R., & Frohlich, D. (2001). Using e-mail for personal relationships. *American Behavioral Scientist, 45*(3), 530–549.

Booth-Kewley, S., Edwards, J. E., & Rosenfeld, P. (1992). Impression management, social desirability, and computer administration of attitude questionnaires: Does the computer make a difference? *Journal of Applied Psychology, 77*(4), 562–566.

Borg, W. R., & Gall, M. D. (1989). *Educational research: An introduction* (5th ed.). New York: Longman.

Borg, W. R., Gall, J. P., & Gall, M. D. (1993). *Applying educational research: A practical guide.* New York: Longman.

Boruch, R. F. (1998). Randomized controlled experiments for evaluation and planning. In L. Bickman & D. J. Rog (Eds.), *Handbook of applied social research methods* (pp. 161–191). Thousand Oaks, CA: Sage.

Bryman, A. (2006). *Mixed methods: A four-volume set.* London: Sage.

Bunge, N. (1985). *Finding the words: Conversations with writers who teach.* Athens, OH: Swallow Press, Ohio University Press.

Cahill, S. E. (1989). Fashioning males and females: Appearance management and the social reproduction of gender. *Symbolic Interaction, 12*(2), 281–298.

Campbell, D., & Stanley, J. (1963). Experimental and quasi-experimental designs for research. In N. L. Gage (Ed.), *Handbook of research on teaching* (pp. 1–76). Chicago: Rand McNally.

Campbell, D. T., & Fiske, D. (1959). Convergent and discriminant validation by the multitrait-multimethod matrix. *Psychological Bulletin, 56*, 81–105.

Campbell, W. G., & Ballou, S. V. (1977). *Form and style: Theses, reports, term papers* (5th ed.). Boston: Houghton Mifflin.

Caracelli, V. J., & Greene, J. C. (1993). Data analysis strategies for mixed-method evaluation designs. *Educational Evaluation and Policy Analysis, 15*(2), 195–207.

Carroll, D. L. (1990). *A manual of writer's tricks.* New York: Paragon.

Carstensen, L. W., Jr. (1989). A fractal analysis of cartographic generalization. *The American Cartographer, 16*(3), 181–189.

Castetter, W. B., & Heisler, R. S. (1977). *Developing and defending a dissertation proposal.* Philadelphia: University of Pennsylvania, Graduate School of Education, Center for Field Studies.

Charmaz, K. (2006). *Constructing grounded theory.* Thousand Oaks, CA: Sage.

Cheek, J. (2004). At the margins? Discourse analysis and qualitative research. *Qualitative Health Research, 14*, 1140–1150.

Cherryholmes, C. H. (1992, August-September). Notes on pragmatism and scientific realism. *Educational Researcher*, 13–17.

Clandinin, D. J., & Connelly, F. M. (2000). *Narrative inquiry: Experience and story in qualitative research.* San Francisco: Jossey-Bass.

Cohen, J. (1977). *Statistical power analysis for the behavioral sciences.* New York: Academic Press.

Cook, T. D., & Campbell, D. T. (1979). *Quasi-experimentation: Design and analysis issues for field settings.* Chicago: Rand McNally.

Cooper, H. (1984). *The integrative research review: A systematic approach.* Beverly Hills, CA: Sage.

Cooper, J. O., Heron, T. E., & Heward, W. L. (1987). *Applied behavior analysis.* Columbus, OH: Merrill.

Corbin, J. M., & Strauss, J. M. (2007). *Basics of qualitative research: Techniques and procedures for developing grounded theory* (3rd ed.). Thousand Oaks, CA: Sage.

Creswell, J. W. (1999) Mixed method research: Introduction and application. In G. J. Cizek (Ed.), *Handbook of educational policy* (pp. 455–472). San Diego, CA: Academic Press.

Creswell, J. W. (2007). *Qualitative inquiry and research design: Choosing among five approaches* (3rd ed.). Thousand Oaks, CA: Sage.

Creswell, J. W. (2008). *Educational research: Planning, conducting, and evaluating quantitative and qualitative research* (3rd ed.). Upper Saddle River, NJ: Merrill.

Creswell, J. W., & Brown, M. L. (1992, Fall). How chairpersons enhance faculty research: A grounded theory study. *The Review of Higher Education, 16*(1), 41–62.

Creswell, J. W., & Miller, D. (2000). Determining validity in qualitative inquiry. *Theory into Practice, 39*(3), 124–130.

Creswell, J. W., & Plano Clark, V. L. (2007). *Designing and conducting mixed methods research.* Thousand Oaks, CA: Sage.

Creswell, J. W., Plano Clark, V., Gutmann, M., & Hanson, W. (2003). Advanced mixed methods designs. In A. Tashakkori & C. Teddlie (Eds.), *Handbook of mixed method research in the social and behavioral sciences* (pp. 209–240). Thousand Oaks, CA: Sage.

Creswell, J. W., Seagren, A., & Henry, T. (1979). Professional development training needs of department chairpersons: A test of the Biglan model. *Planning and Changing, 10,* 224–237.

Crotty, M. (1998). *The foundations of social research: Meaning and perspective in the research process.* London: Sage.

Crutchfield, J. P. (1986). *Locus of control, interpersonal trust, and scholarly productivity.* Unpublished doctoral dissertation, University of Nebraska-Lincoln.

DeGraw, D. G. (1984). *Job motivational factors of educators within adult correctional institutions from various states.* Unpublished doctoral dissertation, University of Nebraska-Lincoln.

Denzin, N. K., & Lincoln, Y. S. (Eds.). (2005). *The handbook of qualitative research* (3rd ed.). Thousand Oaks, CA: Sage.

Dillard, A. (1989). *The writing life.* New York: Harper & Row.

Dillman, D. A. (1978). *Mail and telephone surveys: The total design method.* New York: John Wiley.

Duncan, O. D. (1985). Path analysis: Sociological examples. In H. M. Blalock, Jr. (Ed.), *Causal models in the social sciences* (2nd ed., pp. 55–79). New York. Aldine.

Educational Resources Information Center. (1975). *Thesaurus of ERIC descriptors* (12th ed.). Phoenix, AZ: Oryx.

Eisner, E. W. (1991). *The enlightened eye: Qualitative inquiry and the enhancement of educational practice.* New York: Macmillan.

Elbow, P. (1973). *Writing without teachers.* London: Oxford University Press.

Enns, C. Z., & Hackett, G. (1990). Comparison of feminist and nonfeminist women's reactions to variants of nonsexist and feminist counseling. *Journal of Counseling Psychology, 37*(1), 33–40.

Fay, B. (1987). *Critical social science.* Ithaca, NY: Cornell University Press.

Field, A., & Hole, G. (2003). *How to design and report experiments.* Thousand Oaks, CA; Sage.

Finders, M. J. (1996). Queens and teen zines: Early adolescent females reading their way toward adulthood. *Anthropology and Education Quarterly, 27,* 71–89.

Fink, A. (2002). *The survey kit* (2nd ed.). Thousand Oaks, CA: Sage.

Firestone, W. A. (1987). Meaning in method: The rhetoric of quantitative and qualitative research. *Educational Researcher, 16,* 16–21.

Flick, U. (Ed.). (2007). *The Sage qualitative research kit.* London: Sage.

Flinders, D. J., & Mills, G. E. (Eds.). (1993). *Theory and concepts in qualitative research: Perspectives from the field.* New York: Columbia University, Teachers College Press.

Fowler, F. J. (2002). *Survey research methods* (3rd ed.). Thousand Oaks, CA: Sage.

Franklin, J. (1986). *Writing for story: Craft secrets of dramatic nonfiction by a two-time Pulitzer prize-winner.* New York: Atheneum.

Gamson, J. (2000). Sexualities, queer theory, and qualitative research. In N. K. Denzin & Y. S. Lincoln (Eds.), *Handbook of qualitative research* (pp. 347–365). Thousand Oaks, CA: Sage.

Gibbs, G. R. (2007). Analyzing qualitative data. In U. Flick (Ed.), *The Sage qualitative research kit*. London: Sage.

Giordano, J., O'Reilly, M., Taylor, H., & Dogra, N. (2007). Confidentiality and autonomy: The challenge(s) of offering research participants a choice of disclosing their identity. *Qualitative Health Research, 17*(2), 264–275.

Glesne, C., & Peshkin, A. (1992). *Becoming qualitative researchers: An introduction*. White Plains, NY: Longman.

Gravetter, F. J., & Wallnau, L. B. (2000). *Statistics for the behavioral science* (5th ed.). Belmont, CA: Wadsworth/Thomson.

Greene, J. C. (2007). *Mixed methods in social inquiry*. San Francisco: Jossey-Bass.

Greene, J. C., & Caracelli, V. J. (Eds.). (1997). *Advances in mixed-method evaluation: The challenges and benefits of integrating diverse paradigms*. (New Directions for Evaluation, No. 74). San Francisco: Jossey-Bass.

Greene, J. C., Caracelli, V. J., & Graham, W. F. (1989). Toward a conceptual framework for mixed-method evaluation designs. *Educational Evaluation and Policy Analysis, 11*(3), 255–274.

Guba, E. G. (1990). The alternative paradigm dialog. In E. G. Guba (Ed.), *The paradigm dialog* (pp. 17–30). Newbury Park, CA: Sage.

Guba, E. G., & Lincoln, Y. S. (2005). Paradigmatic controversies, contradictions, and emerging confluences. In N. K. Denzin & Y. S. Lincoln, *The Sage handbook of qualitative research* (3rd ed., pp. 191–215). Thousand Oaks, CA: Sage.

Hatch, J. A. (2002). *Doing qualitative research in educational settings*. Albany: State University of New York Press.

Heron, J., & Reason, P. (1997). A participatory inquiry paradigm. *Qualitative Inquiry, 3,* 274–294.

Hesse-Bieber, S. N., & Leavy, P. (2006). *The practice of qualitative research*. Thousand Oaks, CA: Sage.

Homans, G. C. (1950). *The human group*. New York: Harcourt, Brace.

Hopkins, T. K. (1964). *The exercise of influence in small groups*. Totowa, NJ: Bedmister.

Hopson, R. K., Lucas, K. J., & Peterson, J. A. (2000). HIV/AIDS talk: Implications for prevention intervention and evaluation. In R. K. Hopson (Ed.), *How and why language matters in evaluation*. (New Directions for Evaluation, Number 86). San Francisco: Jossey-Bass.

Hossler, D., & Vesper, N. (1993). An exploratory study of the factors associated with parental savings for postsecondary education. *Journal of Higher Education, 64*(2), 140–165.

Houtz, L. E. (1995). Instructional strategy change and the attitude and achievement of seventh- and eighth-grade science students. *Journal of Research in Science Teaching, 32*(6), 629–648.

Huber, J., & Whelan, K. (1999). A marginal story as a place of possibility: Negotiating self on the professional knowledge landscape. *Teaching and Teacher Education, 15,* 381–396.

Humbley, A. M., & Zumbo, B. D. (1996). A dialectic on validity: Where we have been and where we are going. *The Journal of General Psychology, 123,* 207–215.

Isaac, S., & Michael, W. B. (1981). *Handbook in research and evaluation: A collection of principles, methods, and strategies useful in the planning, design, and evaluation of studies in education and the behavioral sciences* (2nd ed.). San Diego, CA: EdITS.

Isreal, M., & Hay, I. (2006). *Research ethics for social scientists: Between ethical conduct and regulatory compliance*. London: Sage.

Janovec, T. (2001). *Procedural justice in organizations: A literature map*. Unpublished manuscript, University of Nebraska-Lincoln.

Janz, N. K., Zimmerman, M. A., Wren, P. A., Isreal, B. A., Freudenberg, N., & Carter, R. J. (1996). Evaluation of 37 AIDS prevention projects: Successful approaches and barriers to program effectiveness. *Health Education Quarterly, 23*(1), 80–97.

Jick, T. D. (1979, December). Mixing qualitative and quantitative methods: Triangulation in action. *Administrative Science Quarterly, 24,* 602–611.

Johnson, R. B., Onwuegbuzie, A. J., & Turner, L.A. (2007). Toward a definition of mixed methods research. *Journal of Mixed Methods Research, 1*(2), 112–133.

Jungnickel, P. W. (1990). *Workplace correlates and scholarly performance of pharmacy clinical faculty members.* Unpublished manuscript, University of Nebraska-Lincoln.

Kalof, L. (2000). Vulnerability to sexual coercion among college women: A longitudinal study. *Gender Issues, 18*(4), 47–58.

Keeves, J. P. (Ed.). (1988). *Educational research, methodology, and measurement: An international handbook.* Oxford, UK: Pergamon.

Kemmis, S., & Wilkinson, M. (1998). Participatory action research and the study of practice. In B. Atweh, S. Kemmis, & P. Weeks (Eds.), *Action research in practice: Partnerships for social justice in education* (pp. 21–36). New York: Routledge.

Keppel, G. (1991). *Design and analysis: A researcher's handbook* (3rd ed.). Englewood Cliffs, NJ: Prentice Hall.

Kerlinger, F. N. (1979). *Behavioral research: A conceptual approach.* New York: Holt, Rinehart & Winston.

Kline, R. B. (1998). *Principles and practice of structural equation modeling.* New York: Guilford.

Kos, R. (1991). Persistence of reading disabilities: The voices of four middle school students. *American Educational Research Journal, 28*(4), 875–895.

Kushman, J. W. (1992, February). The organizational dynamics of teacher workplace. *Educational Administration Quarterly, 28*(1), 5–42.

Kvale, S. (2007). Doing interviews. In U. Flick (Ed.), *The Sage qualitative research kit.* London: Sage.

Labovitz, S., & Hagedorn, R. (1971). *Introduction to social research.* New York: McGraw-Hill.

Ladson-Billings, G. (2000). Racialized discourses and ethnic epistemologies. In N. K. Denzin & Y. S. Lincoln (Eds.), *Handbook on qualitative research* (pp. 257–277). Thousand Oaks, CA: Sage.

Lather, P. (1986). Research as praxis. *Harvard Educational Review, 56,* 257–277.

Lather, P. (1991). *Getting smart: Feminist research and pedagogy with/in the postmodern.* New York: Routledge.

Lauterbach, S. S. (1993). In another world: A phenomenological perspective and discovery of meaning in mothers' experience with death of a wished-for baby: Doing phenomenology. In P. L. Munhall & C. O. Boyd (Eds.), *Nursing research: A qualitative perspective* (pp. 133–179). New York: National League for Nursing Press.

LeCompte, M. D., & Schensul, J. J. (1999). *Designing and conducting ethnographic research.* Walnut Creek, CA: AltaMira.

Lee, Y. J., & Greene, J. (2007). The predictive validity of an ESL placement test: A mixed methods approach. *Journal of Mixed Methods Research, 1*(4), 366–389.

Leslie, L. L. (1972). Are high response rates essential to valid surveys? *Social Science Research, 1,* 323–334.

Lincoln, Y. S., & Guba, E. G. (1985). *Naturalistic inquiry.* Beverly Hills, CA: Sage.

Lincoln, Y. S., & Guba, E. G. (2000). Paradigmatic controversies, contradictions, and emerging confluences. In Y. S. Lincoln & E. G. Guba (Eds.), *Handbook of qualitative research* (pp. 163–188). Thousand Oaks, CA: Sage.

Lipsey, M. W. (1990). *Design sensitivity: Statistical power for experimental research.* Newbury Park, CA: Sage.

Locke, L. F., Spirduso, W. W., & Silverman, S. J. (2007). *Proposals that work: A guide for planning dissertations and grant proposals* (5th ed.). Thousand Oaks, CA: Sage.

Lysack, C. L., & Krefting, L. (1994). Qualitative methods in field research: An Indonesian experience in community based practice. *The Occupational Therapy Journal of Research, 14*(20), 93–110.

Marshall, C., & Rossman, G. B. (2006). *Designing qualitative research* (4th ed.). Thousand Oaks, CA: Sage.

Mascarenhas, B. (1989). Domains of state-owned, privately held, and publicly traded firms in international competition. *Administrative Science Quarterly, 34*, 582–597.

Maxwell, J. A. (2005). *Qualitative research design: An interactive approach* (2nd ed.). Thousand Oaks, CA: Sage.

McCracken, G. (1988). *The long interview.* Newbury Park, CA: Sage.

Megel, M. E., Langston, N. F., & Creswell, J. W. (1987). Scholarly productivity: A survey of nursing faculty researchers. *Journal of Professional Nursing, 4*, 45–54.

Merriam, S. B. (1998). *Qualitative research and case study applications in education.* San Francisco: Jossey-Bass.

Mertens, D. M. (1998). *Research methods in education and psychology: Integrating diversity with quantitative and qualitative approaches.* Thousand Oaks, CA: Sage.

Mertens, D. M. (2003). Mixed methods and the politics of human research: The transformative-emancipatory perspective. In A. Tashakkori & C. Teddlie (Eds.), *Handbook of mixed methods in the social & behavioral sciences* (pp. 135–164). Thousand Oaks, CA: Sage.

Miles, M. B., & Huberman, A. M. (1994). *Qualitative data analysis: A sourcebook of new methods.* Thousand Oaks, CA: Sage.

Miller, D. (1992). *The experiences of a first-year college president: An ethnography.* Unpublished doctoral dissertation, University of Nebraska-Lincoln.

Miller, D. C. (1991). *Handbook of research design and social measurement* (5th ed.). Newbury Park, CA: Sage.

Moore, D. (2000). Gender identity, nationalism, and social action among Jewish and Arab women in Israel: Redefining the social order? *Gender Issues, 18*(2), 3–28.

Morgan, D. (1998). Practical strategies for combining qualitative and quantitative methods: Applications to health research. *Qualitative Health Research, 8*(3), 362–376.

Morgan, D. (2007). Paradigms lost and pragmatism regained: Methodological implications of combining qualitative and quantitative methods. *Journal of Mixed Methods Research, 1*(1), 48–76.

Morse, J. M. (1991, March/April). Approaches to qualitative-quantitative methodological triangulation. *Nursing Research, 40*(1), 120–123.

Morse, J. M. (1994). Designing funded qualitative research. In N. K. Denzin & Y. S. Lincoln (Eds.), *Handbook of qualitative research* (pp. 220–235). Thousand Oaks, CA: Sage.

Moustakas, C. (1994). *Phenomenological research methods.* Thousand Oaks, CA: Sage.

Murguia, E., Padilla, R. V., & Pavel, M. (1991, September). Ethnicity and the concept of social integration in Tinto's model of institutional departure. *Journal of College Student Development, 32*, 433–439.

Murphy, J. P. (1990). *Pragmatism: From Peirce to Davidson.* Boulder, CO: Westview.

Nesbary, D. K. (2000). *Survey research and the world wide web.* Boston: Allyn & Bacon.

Neuman, S. B., & McCormick, S. (Eds.). (1995). *Single-subject experimental research: Applications for literacy.* Newark, DE: International Reading Association.

Neuman, W. L. (2000). *Social research methods: Qualitative and quantitative approaches* (4th ed.). Boston: Allyn & Bacon.

Newman, I., & Benz, C. R. (1998). *Qualitative-quantitative research methodology: Exploring the interactive continuum.* Carbondale and Edwardsville: Southern Illinois University Press.

Nieswiadomy, R. M. (1993). *Foundations of nursing research.* (2nd ed.). Norwalk, CT: Appleton & Lange.

Nordenmark, M., & Nyman, C. (2003). Fair or unfair? Perceived fairness of household division of labour and gender equality among women and men. *The European Journal of Women's Studies, 10*(2), 181–209.

O'Cathain, A., Murphy, E., & Nicholl, J. (2007). Integration and publications as indicators of "yield" from mixed methods studies. *Journal of Mixed Methods Research, 1*(2), 147–163.

Olesen, V. L. (2000). Feminism and qualitative research at and into the millennium. In N. L. Denzin & Y. S. Lincoln, *Handbook of qualitative research* (pp. 215–255). Thousand Oaks, CA: Sage.

Onwuegbuzie, A. J., & Johnson, R. B. (2006). The validity issue in mixed research. *Research in the Schools, 13*(1), 48–63.

Padula, M. A., & Miller, D. (1999). Understanding graduate women's reentry experiences. *Psychology of Women Quarterly, 23,* 327–343.

Patton, M. Q. (1990). *Qualitative evaluation and research methods* (2nd ed.). Newbury Park, CA: Sage.

Patton, M. Q. (2002). *Qualitative research and evaluation methods* (3rd ed.). Thousand Oaks, CA: Sage.

Phillips, D. C., & Burbules, N. C. (2000). *Postpositivism and educational research.* Lanham, NY: Rowman & Littlefield.

Plano Clark, V. L., & Creswell, J. W. (2008). *The mixed methods reader.* Thousand Oaks, CA: Sage.

Prose, F. (2006). *Reading like a writer.* New York: HarperCollins.

Punch, K. F. (2000). *Developing effective research proposals.* London: Sage.

Punch, K. F. (2005). *Introduction to social research: Quantitative and qualitative approaches* (2nd ed.). London: Sage.

Reichardt, C. S., & Mark, M. M. (1998). Quasi-experimentation. In L. Bickman & D. J. Rog (Eds.), *Handbook of applied social research methods* (pp. 193–228). Thousand Oaks, CA: Sage.

Rhoads, R. A. (1997). Implications of the growing visibility of gay and bisexual male students on campus. *NASPA Journal, 34*(4), 275–286.

Richardson, L. (1990). *Writing strategies: Reaching diverse audiences.* Newbury Park, CA: Sage.

Richie, B. S., Fassinger, R. E., Linn, S. G., Johnson, J., Prosser, J., & Robinson, S. (1997). Persistence, connection, and passion: A qualitative study of the career development of highly achieving African American-Black and White women. *Journal of Counseling Psychology, 44*(2), 133–148.

Riemen, D. J. (1986). The essential structure of a caring interaction: Doing phenomenology. In P. M. Munhall & C. J. Oiler (Eds.), *Nursing research: A qualitative perspective* (pp. 85–105). Norwalk, CN: Appleton-Century-Crofts.

Rogers, A., Day, J., Randall, F., & Bentall, R. P. (2003). Patients' understanding and participation in a trial designed to improve the management of anti-psychotic medication: A qualitative study. *Social Psychiatry and Psychiatric Epidemiology, 38,* 720–727.

Rorty, R. (1983). *Consequences of pragmatism.* Minneapolis: University of Minnesota Press.

Rorty, R. (1990). Pragmatism as anti-representationalism. In J. P. Murphy, *Pragmatism: From Peirce to Davison* (pp. 1–6). Boulder, CO: Westview.

Rosenthal, R., & Rosnow, R. L. (1991). *Essentials of behavioral research: Methods and data analysis.* New York: McGraw-Hill.

Ross-Larson, B. (1982). *Edit yourself: A manual for everyone who works with words.* New York: Norton.

Rossman, G., & Rallis, S. F. (1998). *Learning in the field: An introduction to qualitative research.* Thousand Oaks, CA: Sage.

Rossman, G. B., & Wilson, B. L. (1985, October). Numbers and words: Combining quantitative and qualitative methods in a single large-scale evaluation study. *Evaluation Review, 9*(5), 627–643.

Rudestam, K. E., & Newton, R. R. (2007). *Surviving your dissertation* (3rd ed.). Thousand Oaks, CA: Sage.

Salant, P., & Dillman, D. A. (1994). *How to conduct your own survey.* New York: John Wiley.

Salkind, N. (1990). *Exploring research.* New York: MacMillan.

Sarantakos, S. (2005). *Social research* (3rd ed.). New York: Palgrave Macmillan.

Schwandt, T. A. (1993). Theory for the moral sciences: Crisis of identity and purpose. In D. J. Flinders & G. E. Mills (Eds.), *Theory and concepts in qualitative research: Perspectives from the field* (pp. 5–23). New York: Columbia University, Teachers College Press.

Schwandt, T. A. (2000). Three epistemological stances for qualitative inquiry. In N. K. Denzin & Y. S. Lincoln (Eds.), *Handbook of qualitative research* (2nd ed., pp. 189–213). Thousand Oaks, CA: Sage.

Schwandt, T. A. (2007). *Dictionary of qualitative inquiry* (3rd ed.). Thousand Oaks, CA: Sage.

Shadish, W. R., Cook, T. D., & Campbell, D. T. (2001). *Experimental and quasi-experimental designs for generalized causal inference.* Boston: Houghton Mifflin.

Sieber, J. E. (1998). Planning ethically responsible research. In L. Bickman & D. J. Rog (Eds.), *Handbook of applied social research methods* (pp. 127–156). Thousand Oaks, CA: Sage.

Sieber, S. D. (1973). The integration of field work and survey methods. *American Journal of Sociology, 78,* 1335–1359.

Slife, B. D., & Williams, R. N. (1995). *What's behind the research? Discovering hidden assumptions in the behavioral sciences.* Thousand Oaks, CA: Sage.

Smith, J. K. (1983, March). Quantitative versus qualitative research: An attempt to clarify the issue. *Educational Researcher,* 6–13.

Spradley, J. P. (1980). *Participant observation.* New York: Holt, Rinehart & Winston.

Stake, R. E. (1995). *The art of case study research.* Thousand Oaks, CA: Sage.

Steckler, A., McLeroy, K. R., Goodman, R. M., Bird, S. T., & McCormick, L. (1992). Toward integrating qualitative and quantitative methods: An introduction. *Health Education Quarterly, 19*(1), 1–8.

Steinbeck, J. (1969). *Journal of a novel: The East of Eden letters.* New York: Viking.

Strauss, A., & Corbin, J. (1990). *Basics of qualitative research: Grounded theory procedures and techniques* (1st ed.). Newbury Park, CA: Sage.

Strauss, A., & Corbin, J. (1998). *Basics of qualitative research: Grounded theory procedures and techniques* (2nd ed.). Thousand Oaks, CA: Sage.

Sudduth, A. G. (1992). *Rural hospitals use of strategic adaptation in a changing health care environment.* Unpublished doctoral dissertation, University of Nebraska-Lincoln.

Sue, V. M., & Ritter, L. A. (2007). *Conducting online surveys.* Thousand Oaks, CA: Sage.

Tarshis, B. (1982). *How to write like a pro: A guide to effective nonfiction writing.* New York: New American Library.

Tashakkori, A., & Creswell, J. W. (2007). Exploring the nature of research questions in mixed methods research. Editorial. *Journal of Mixed Methods Research, 1*(3), 207–211.

Tashakkori, A., & Teddlie, C. (1998). *Mixed methodology: Combining qualitative and quantitative approaches.* Thousand Oaks, CA: Sage.

Tashakkori, A., & Teddlie, C. (Eds.). (2003). *Handbook of mixed method research in the social and behavior sciences.* Thousand Oaks, CA: Sage.

Teddlie, C., & Yu, F. (2007). Mixed methods sampling: A typology with examples. *Journal of Mixed Methods Research, 1*(1), 77–100.

Terenzini, P. T., Cabrera, A. F., Colbeck, C. L., Bjorklund, S. A., & Parente, J. M. (2001). Racial and ethnic diversity in the classroom. *The Journal of Higher Education, 72*(5), 509–531.

Tesch, R. (1990). *Qualitative research: Analysis types and software tools.* New York: Falmer.

Thomas, G. (1997). What's the use of theory? *Harvard Educational Review, 67*(1), 75–104.

Thomas, J. (1993). *Doing critical ethnography.* Newbury Park, CA: Sage.

Thorndike, R. M. (1997). *Measurement and evaluation in psychology and education* (6th ed.). New York: Macmillan.

Trujillo, N. (1992). Interpreting (the work and the talk of) baseball: Perspectives on ballpark culture. *Western Journal of Communication, 56,* 350–371.

Tuckman, B. W. (1999). *Conducting educational research* (5th ed.). Fort Worth, TX: Harcourt, Brace.

Turabian, K. L. (1973). *A manual for writers of term papers, theses, and dissertations* (4th ed.) Chicago: University of Chicago Press.

University of Chicago Press. *A manual of style.* (1982). Chicago: Author.

University Microfilms. (1938–). *Dissertation abstracts international.* Ann Arbor, MI: Author.

VanHorn-Grassmeyer, K. (1998). *Enhancing practice: New professional in student affairs.* Unpublished doctoral dissertation, University of Nebraska-Lincoln.

Van Maanen, J. (1988). *Tales of the field: On writing ethnography.* Chicago: University of Chicago Press.

Vernon, J. E. (1992). *The impact of divorce on the grandparent/grandchild relationship when the parent generation divorces.* Unpublished doctoral dissertation, University of Nebraska-Lincoln.

Vogt, W. P. (1999). *Dictionary of statistics and methodology: A nontechnical guide for the social sciences* (2nd ed.). Thousand Oaks, CA: Sage.

Webb, R. B., & Glesne, C. (1992). Teaching qualitative research. In M. D. LeCompte, W. L. Millroy, & J. Preissle (Eds.), *The Handbook of qualitative research in education* (pp. 771–814). San Diego, CA: Academic Press.

Webb, W. H., Beals, A. R., & White, C. M. (1986). *Sources of information in the social sciences: A guide to the literature* (3rd ed.). Chicago: American Library Association.

Weitzman, E. A., & Miles, M. B. (1995). *Computer programs for qualitative data analysis.* Thousand Oaks, CA: Sage.

Weitzman, P. F., & Levkoff, S. E. (2000). Combining qualitative and quantitative methods in health research with minority elders: Lessons from a study of dementia caregiving. *Field Methods, 12*(3), 195–208.

Wilkinson, A. M. (1991). *The scientist's handbook for writing papers and dissertations.* Englewood Cliffs, NJ: Prentice Hall.

Wolcott, H. T. (1994). *Transforming qualitative data: Description, analysis, and interpretation.* Thousand Oaks, CA: Sage.

Wolcott, H. T. (1999). *Ethnography: A way of seeing.* Walnut Creek, CA: AltaMira.

Wolcott, H. T. (2001). *Writing up qualitative research* (2nd ed.). Thousand Oaks, CA: Sage.

Yin, R. K. (2003). *Case study research: Design and methods* (2nd ed.). Thousand Oaks, CA: Sage.

Ziller, R. C. (1990). *Photographing the self: Methods for observing personal orientations.* Newbury Park, CA: Sage.

Zinsser, W. (1983). *Writing with a word processor.* New York: Harper Colophon.

Author Index

Subject Index

Note: In page references, b indicates boxes, f indicates figures, and t indicates tables.

TO THE OWNER OF THIS BOOK:

We hope that you found John W. Creswell's **Research Design, 3rd Edition**, helpful. We would appreciate it if you could take a few moments to respond to the following questions. The author and the publisher take all comments into account when planning for any new edition, so we hope you will take the time to return the survey. You can also access these same questions online at www.sagepub.com/creswellsurvey, if that will make it easier for you. Thank you.

Your school/organization: _____

Name of the course: _____

Department: _____ Professor/instructor's name: _____

1. What I found the most valuable about this book: _____

2. Did you experience any problems with the presentation of specific content in the book? Please be specific:
   _____

3. What is your overall impression of this book?
   _____

4. Were you assigned all the chapters to read? If not, what chapters were not assigned?
   _____

5. Did you use the "Writing Exercises"?
   _____

6. Did you buy this book new____ or used ____? Did you buy at the bookstore or through an online retailer? _____

7. Do you plan to keep this book? Yes _____ No _____

8. If you used this book in a course, do you think it helped your final grade?
   Yes ____ No _____

9. In the space below or on a separate sheet of paper, please provide any suggestions you have for improvements the next time this book is revised.
   _____
   _____
   _____

(Optional) Your Name_____ Date _____

May we quote you in the promotion of *Research Design* by John W. Creswell?
Yes _____No_____

2. Fold here, and TAPE shut. Post office will not deliver if stapled.

BUSINESS REPLY MAIL
FIRST-CLASS MAIL   PERMIT NO. 20   THOUSAND OAKS, CA

POSTAGE WILL BE PAID BY ADDRESSEE

Attention: Vicki Knight, Senior Editor
SAGE PUBLICATIONS
PO BOX 5084
THOUSAND OAKS CA 91359-9702

1. Fold here.